CAPE ISLAND

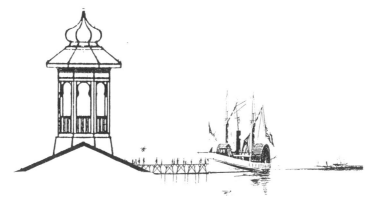

by I. Yevish

A NOVEL OF HISTORIC CAPE MAY

For Ruth and Gloria,
More than sisters—
Friends.

Library of Congress Catalog Card Number
90-93097

ISBN: 0-9626330-1-1

Published by
I. Yevish Books
Box 366
Cape May Point, New Jersey 08212

Printed in the United States of America

1846

CAPTAIN IRONS

SHIPWRECK

It was Willacassa who first saw the ship in distress. The storm had been howling through the shutters and crevices in the old house, awakening the child. And when she went to the window on the oceanside, where the shutters had been removed for repair, she saw it and immediately called for her father in the next room. Harrah came running and joined Willacassa at the window. Even through the rain and mist and the gray night, he could see the ship was in deep trouble.

A tall schooner it was, two-masted, pointing towards shore and plunging into the trough of a huge wave. Its jib sail was flying loose. Its topsails had been carried away, leaving only the naked yards with their flapping tails of stays and guy ropes. And there was doubt in Harrah's mind whether the rigging of the mainmast would survive the strain of water and slanting wind.

"Look how close it is," shouted Willacassa. "Do you think it will crash through our window?"

In spite of the seriousness of the situation, Harrah laughed. "I think not, Cassie. Not just yet anyway."

"But it seems so close!"

The girl was not exaggerating. The ship had drawn so close to the beach that the crew could be seen struggling with the lines. It was too close in fact for life boats to be lowered. Besides, the sea was too tempestuous for open boats, pouring across the top deck as though it were dashing out a fire. And with each surge it washed a man overboard.

"I've got to get down there! Stay here, Cassie. Wake Jonathan and Rachel if you're frightened." Pulling on his trousers, Harrah hurried downstairs and out the door.

From the window Willacassa saw her father emerge from the house and dash towards the water's edge. He was joined by several occupants of the Columbia Hotel and a couple of stretcher bearers, running with slickers over their heads.

It now began to rain heavily. It was impossible to tell where the ocean ended and the flood of rain began. But it was clear—with its naked hull exposed to the keel—the troubled schooner would soon be beached. Tearing a path through the wet sand, it slid to a stop and toppled over on its side. Those crew members clinging to the deck were now flung free. Waist-deep in water, they scampered to safety before the main topmast came crashing down after them.

Harrah and the others lent them a hand. They shouldered the

exhausted crewmen away from the strand. Then the doors of the Columbia Hotel were thrown open and the seamen disappeared into the yellow light of the downstairs kitchen.

When Harrah returned to the wrecked schooner for a last look, he discovered another mariner making for shore. Wading into the sea, Harrah pulled the man to high ground.

"Thank you, Lad. I didn't think I was going to make it."

Even as he talked, water poured out of the man's mouth and ears. It was clear he had swallowed more than a tin cup's worth of ocean. Though he was soaked through and his jacket was washed half off his shoulders, one could tell from the man's buttons and his coat and his white linen shirt that the survivor was a ship's officer.

"Captain Irons, late of Newport," the man said, introducing himself as he lay spent on the beach. Undaunted by the driving rain and the wet sand, he leaned on his arm and twisted his neck in the direction of his wrecked ship. "And that," he declared, letting more water run out of his mouth, "is the *Carolina*."

Harrah looked in the direction of the ship, which lay like a wounded bird on the beach, its wings crushed, its feathers frayed and water-logged.

"There's no help for her," said Harrah. "But we'd best get you into the house."

"Can't move just yet. The wind's knocked out of me."

"Then I'll carry you," said Harrah. "Brace your arms round my neck as I raise you up."

The captain did as he was told, and Harrah struggled with the man, trudging up the soggy beach to the house where Willacassa was waiting for them at the door.

Captain Irons made a quick recovery from his exertions of the night before. By noon he was sitting up in his bed, a cheat quilt at his legs, in one of the guest rooms where Willacassa carried him a breakfast tray. Rachel had prepared a mess of bacon and eggs and some oatmeal bread in the kitchen, but Willacassa insisted on bringing the tray upstairs herself. Even Jonathan, who had slept through the night's events but was anxious to see the sea captain, was unable to dissuade her.

"Ah, the young lady of the house. And how are you this morning, Willa—"

"Willacassa, Captain Irons. My brother wanted to wait on you, but I told him you preferred to be served by me. That's true, isn't it?"

"How could it be otherwise?"

"You look like Captain Whilldin," observed Willacassa, setting the tray down, "only younger."

"Captain Whilldin?"

"He's a steamboat captain who lives on Cape Island. We have a lot of steamboat pilots living here."

"Is that so?" Captain Irons looked depressed even as he tucked a napkin under his chin. "I'm afraid I'll be a captain without a ship for a while. I'll have to go back to Newport and find another."

"Your ship is a total wreck," agreed Willacassa who had been to the beach that morning to check things out. "It's all broken up and lying on its side."

"The *Carolina* was a fine vessel. She'll be sorely missed," announced Captain Irons, more for his own edification than Willacassa's.

"You should know better than to make for Cape May in a storm."

Captain Irons nodded between bites. "I was hoping to put ashore in Delaware. The squall came all of a sudden without proper warning." He leaned back on his pillows as if to give his anxieties a rest.

"You'll sail another," smiled Willacassa. "We have ships coming and going all the time."

"You're a real comfort, Willa—"

"Willacassa."

"And a pretty girl at that."

Willacassa's father and her brother Jonathan appeared at the bedroom door to see the patient. To Harrah, Captain Irons looked as upstanding a New Englander as ever sailed a ship. His eyes were teal blue and his hair puritan dark with just enough gray at the temples to bend its strictness. There was little trace of last night's ordeal on his face. Only a crinkling of the skin near the eyes and a slight scratch on his chin. If he had a weakness it was in the mouth, which was almost no mouth at all.

"How are you, Mr. Harrah?" Captain Irons extended a gold-ringed hand. "I have much to thank you for."

"It's the people of the Columbia Hotel deserve thanks, not myself," said Harrah. "Every member of the crew is accounted for and well. Can't say the same for your cargo. The hold broke open and the casks were lost at sea. What were you carrying?"

"Molasses. And hogsheads of brown sugar from Jamaica. Not a very rich cargo. But a loss just the same. The schooner was the real loss. Nothing can be salvaged, I suppose?"

"Not that I can see."

"Well, at least I'm in one piece," remarked the captain. "And in very good hands. Do you think my men and I can get passage to New York?"

"Maybe the *Palmetto* can take you."

"The *Palmetto?*"

"Yes, the ship's the guest of the Atlantic Hotel. I believe it's leaving for New York tomorrow. It too was shook up by the storm. But it's still seaworthy. If you like, I'll speak to Captain Christianson?"

"Much obliged," nodded Irons. "You see, I'm anxious to get back to my son. Before you know it, rumor'll have it that I was lost at sea."

Captain Irons remained in bed the rest of the morning. But at noon he appeared in the dining room, dressed in his uniform which Rachel had washed and ironed for him. He was quite taciturn when Harrah asked him about his health, replying tersely, "Fine. Just fine."

But when Willacassa ran up to him and put her arms around his neck, he dropped his reserve.

"You've been walking on the beach, I see."

"You see? How did you know that?" she asked.

"I watched you from the window."

"With a spyglass"

"No, I lost that with my ship."

"I was looking for treasure," said Willacassa.

"Treasure?"

"All ships carry treasure, don't they?"

"Not the *Carolina*. But it's possible my spyglass washed ashore."

Willacassa shook her head in contradiction. "No, not your spyglass. Something else."

"Something else?" The face of Captain Irons darkened.

"Are you sure there's no treasure on your ship?"

Captain Irons raised his right hand. "Word of honor."

Willacassa then pulled something from her pocket. "Then what's the key for?"

"The key?" Captain Irons went from sober to a broad smile. "Well, if you aren't my ship's angel! So it is a key. The key to the captain's chest." He tucked into his pocket and pulled out a gold piece which he gave to her. "Just as surely as the lighthouse beckoned me to shore, so this pretty little lass showed me the key to my ship's chest. Safe to say, however, that it's not worth unlocking—even if it were found. On my next boat, Willacassa, I'll make a figurehead of you. And I'll call it 'Angel of Cape Island.'

For an angel you are. And I the angel's benefactor.''

Her father was amused by the captain's hyperbolic exclamations, for in the presence of anyone but Willacassa, Captain Irons was strangely close-mouthed, the very picture of a tight-lipped New Englander. But Harrah did not fully understand why the man was so sanguine. True, he had seen his crew rescued; not a man was missing. But he had lost his ship and his ship's cargo. And no ship's company, however heavily insured, was satisfied with a claim when a safe arrival was called for. It was almost as if Captain Irons had been lifted of a weighty responsibility, as if he had with Willacassa's discovery been freed of some troubling concern.

"That was no key to a sea chest," remarked Jonathan later on.

"No? Why not?" Harrah found his son full of useless information.

"Too big. And it's made of iron, not brass.''

"But his name is Captain Irons.''

That was true enough. Still his son was not amused.

"Do you think I'll ever make captain of a ship?'' asked Jonathan.

"If you want to, Son. And work hard at it. But you've got to start as a seaman first.''

"I don't want to be a seaman," said the eleven-year-old Jonathan. "They don't wear gold rings on their fingers. I want to run a ship from the start, like Captain Irons.''

A wise young man, thought Harrah. But what was happening to the old values, the old creeds? Didn't youth earn its way up the ladder anymore? Whatever became of the Puritan work ethic Harrah had known as a boy? Obviously he was doing something wrong as Jonathan's father. He was indulging the lad, just as one indulged a storyteller his fictions. Still he laughed at the idea, just as he was amused by Willacassa's precocity when that manifested itself. But even as he laughed he was somewhat troubled. Captain Irons had made an impact, however brief his stay. Harrah hoped it was not a permanent one.

HIGBEE'S BEACH

Harrah believed that personality was ingrained from birth. He found strong evidence for this in his daughter. Willacassa had a will and determination that no training or circumstance could account for. And she had definite ideas about what she wanted to do and when she wanted to do it, and would brook no interference.

Though she had the broad strand to run on in either direction, Willacassa preferred the rough, deep sand and dunes of Higbee's Beach for an outing. The strand which lay at the foot of Trescott House would always be there for Willacassa, its waves crawling up the white beach, like fingers on one of Rachel's apple pies. But Higbee's Beach, some three miles away, was something special. One could cook out there among the dunes and find Cape May Diamonds and watch the horseshoe crabs mate. And after a storm, the flotsam and debris of the Landing would wash ashore, making curious studies and good pickings for a child who had only dolls and doll furniture to amuse her.

Unable to persuade her brother Jonathan to come along—"Who wants to spend a day with Willacassa picking pebbles!"—Harrah hitched up the wagon and helped his daughter to her seat. Across the road stood the overwhelming, new Columbia House that George Hildreth had built, and all Harrah could do was shake his head at the sight of it. Mounting the wagon, he chucked his horse up Ocean Street, turning left on Washington Avenue where, at the corner of Jackson, Willacassa saw a group of men standing outside of Riddle Tavern. Coming at last upon huge, sprawling Congress Hall, they swung away from the landmark, turning right on Perry Street.

After crossing the wooden bridge over Cape Island Creek, Harrah quickened the pace, driving his horse past the cluster of Eldridge homes, the smith shop, the ox shop, the Cox houses (among others), and the school house.

Willacassa always made a face when she rode past the schoolhouse and today was no different. She knew the school door would be open soon and with that would come an end to her freedom. All she ever liked about the school were the brief fifteen-minute recesses and the meting out of punishment to the boys, the whippings or the shutting up in the closet and the pouring of cold water down their backs. (The boys were fresh and deserved everything they got, even if they got it for the wrong reasons.) By making a face, Willacassa felt she was somehow putting off the day when she had to sit on a hard wooden bench with a slate in her hand from eight in the morning till noon and from one to four in the afternoon, even if the calendar told her otherwise.

Harrah tied up at Captain Higbee's Hermitage Hotel and allowed Willacassa to run the rest of the way to the beach while he followed her with provisions and a frying pan. He marvelled how the girl was able to fly over the sandy path while he seemed to sink deeper with each step. But then again he marvelled at everything Willacassa did. Not yet nine, she was already an independent creature, as strong-willed and as unorthodox as her mother had been. And she was at least as pretty.

The path Harrah followed was thickly wooded on both sides, a tangle of cedar and windswept junipers. The bay lay ahead just out of sight. He expected to find Willacassa at the shoreline, her back to the bay, all smiles at having gotten there first, or with a catch of Cape May Diamonds in her palms.

But before he reached the beach, Harrah saw her come running towards him.

"Don't look, Daddy," the girl cried hoarsely. Her face was wraithlike and forbidding.

"Don't look at what?"

"You won't like what you see." She tried to push him back with open hands.

"Come now, Willacassa. You don't expect me to trudge all this way and not—"

Harrah took several steps forward, then stopped abruptly. Directly ahead, at the edge of the beach, half in water lay what Willacassa did not want him to see. Six blacks, their faces bloated with sea water, washed ashore like jellyfish, were sprawled lifeless on the sand. Tied to one another with chains, they were a hopeless tangle of limbs and twisted bodies. Death had caught the six in their final agony, arms thrust out and legs splayed in a desperate attempt to stay afloat. The men apparently had been thrown overboard during the storm. No sea captain wants an illegal cargo exposed in a shipwreck. But to toss them overboard like so many bales of cotton! Harrah quickly fell sick at the idea.

"Who are they?" asked Willacassa in a whisper. It was clear she was on the verge of hysteria, but she clung fiercely to Harrah's arm.

"Slaves. Or at least they were to be sold as slaves."

"Slaves? Sold where?" Willacassa had heard of slaves, had even seen black servants and housemaids on the Island, but she never really grasped what they were all about.

"In the South. Georgia, I suppose. Or the Carolinas."

"The name of the wrecked ship was 'Carolina,' wasn't it,

Daddy? Captain Irons's ship?"

"Yes," repeated Harrah, leading the girl away. "Captain Irons's ship."

1847

GEORGINA BROOKENS

OLD ACQUAINTANCE

Harry of the West was coming, and the whole island was gearing up to greet him. Reverend Moses Williamson was to be the principal speaker at a convocation held in his honor. And the town was expected to be filled to overflowing.

But Nathaniel Harrah had other things on his mind. For a few years he had run a modest but reasonably successful guest house. The food was good, the rates cheap, and the rooms comfortable. If not exactly elegant, Trescott House at least enabled Harrah to put bread on the table, next to the chicken or fish platters prepared by Rachel for his children.

Then came the new Columbia House across the street. An imposing structure with high-pillared verandas extending round the entire house—Harrah never failed to marvel at how pillars so slender and airy could sustain the porches—the hotel had some ninety feet of frontage on the west side of Ocean Street. And with its new wing completed this very year, an addition four stories high, Columbia House all but dwarfed the tiny Shields place and Harrah's own Trescott House.

More than that, pomp and formality were the rule at Captain George Hildreth's establishment. Dinner was taken at three o'clock instead of half past one in the huge dining hall. And the tables were set with Irish linen, sterling silver, and decanters of burgundy and madeira. The waiters, colored and white, were under the supervison of the aristocratic Mr. Lilburn Harwood.

As a result, guests at the Trescott House were in short supply. Visitors to Cape Island had money to spend and they might as well spend it in comparative luxury. Not only were the walls of his house closing in on Harrah, the lamp of better days was growing dim. Though not by nature a worrier, he was troubled this time. With the mirror of failure staring him in the face, he saw his situation as desperate.

Oh, he still had savings to draw on. But he was feeling the squeeze, and the prospect of ruin loomed large. How did a man stave off poverty? How did a man protect his children from the biting pinch of hunger? How did a man deal with a condition that would leave his loved ones at the mercy of the elements? How did a man reconcile himself to the fact that he was not a good provider? Not a breadwinner? Not worthy of the memory of his dead wife?

He could, of course, move on and get a job somewhere. Harrah was still strong and not without experience. But he had put the gypsy life behind him, and he wanted no more of wandering from place to place. He had searched for a proper setting for his children and now that he had found one on Cape Island he was not about to give it up.

All these things preoccupied Harrah as he paced up and down Ocean Street, looking for some way to improve his fortunes.

"I thought it was you, Nathaniel. But I couldn't be sure. Last time I saw you, you were an ironmaster in the Pines."

Harrah had no trouble recognizing Georgina Pierot, still beautiful after these many years, if somewhat fuller of figure.

"Georgina! I had no idea you were at Cape Island."

When she leaned towards him for an embrace, he kissed her on the cheek.

"You can do better than that, Nathaniel Harrah. But seeing we're on the street, I'll forgive you."

"How have you been?" asked Harrah who had last met her at Quaker Bridge in the Pines. From all appearances Georgina was quite well, as becoming as ever in sky blue.

"The same as before. Only a widow in the making."

"A widow?"

"Yes, I've been married to George Brookens. You remember, my step-brother—the railroadman. But poor George is breathing his last. Has been consumptive for some time now. My first husband, Gerald, gambled himself out of existence. And I'm afraid this one will cough himself to death. I suppose I should have steered clear of railroad men. But all that is spilt milk now."

"Will you be staying for a while at Cape Island?"

"Until the end, the bitter end. Which shouldn't be far off. Sea air helps George breathe. But he hasn't lung enough to do it with."

Harrah could easily understand Georgina's impatience. Her lot was not an easy one. But he could not reconcile her ready acceptance of oncoming widowhood.

"Aren't you going to tell me where you're staying?" she asked mischievously.

"I *live* on Cape Island," said Harrah. "With my two children and a cook."

"And Mrs. Harrah?"

"Dead these past six years."

"I'm sorry."

Georgina took his arm. "I would love to see you while down here. Nathaniel. Could you find some time for me?"

"Actually, it's my busy season. I keep a guest house."

"Trescott House?"

Harrah nodded.

"I thought the name rang familiar. It was Trescott where you were ironmaster, wasn't it? Somewhere in the Pines?"

"You have a good memory."

"Indeed I do—particularly where you're concerned. I was extremely fond of you, you know." She squeezed his arm. "But I see I'm causing you embarrassment. Still, I do want to see you— with George or alone. Is it possible?"

"Of course, Georgina. But if your husband is as ill as you say, it would be best that I visit you. Some time after dinner, perhaps."

"Fine, Nathaniel. I'll wait on the veranda. We face the ocean. Just ask the clerk where the room is. If George is cranky, pay him no mind. It's the privilege of the dying to be cranky."

Harrah was relieved that Georgina was not coming to Trescott House. Though she was an old acquaintance, he did not want her to see that his business was doing poorly. In the old days he had been an ironmaster, and before that the proprietor of a very successful tavern at Cooper's Point. Now things were on the downturn. Although it was August, Trescott House was only half filled.

Besides, there was the problem of his daughter. All women in their thirties set up in Willacassa's mind a host of romantic possibilities. Wasn't her father an attractive man? Didn't he need companionship? But there was contradiction in her. In the end Harrah would have to send the ladies packing. Otherwise it would not be possible for Willacassa to be alone with him again, to claim his attention, and to enjoy once more his undivided adoration.

Towards six o'clock Harrah climbed upstairs the Columbia House to the designated room and found Georgina and her stricken husband on the veranda. George Brookens had slipped badly over the years. Never a handsome man, he displayed a certain stolidity or steadfastness in the past that was now totally absent. Covered with a blanket on this rather cool evening, he sat in his chair with his chin on his chest and his eyes fixed at the ocean, a handkerchief clutched in his hand.

"I'm afraid George is not too well this evening. But he do remember you and wonders whether the rumor that Henry Cl will be coming to Cape Island is a true one."

"It's true," said Harrah. "It's reported that Mr. Clay lost a in the Mexican War. He's coming to Cape Island at the invita of Richard Ludlum. And, of course, he'll stay at the Mansion H for solace and rest."

"We'll all be getting solace and rest soon enough," rem Brookens with a grim smile.

"George is no longer with the Camden and Amboy," a Georgina. "He's sold his shares—at a handsome profit, t add."

"Will make you a rich widow," muttered Brookens, trying to cough but unable to.

"Nonsense, George! You're showing improvement all the time."

"This time last year I was able to get about. Now I'm glued to the bed or chair."

"Getting back to Mr. Clay," said Harrah. "I hear that people are coming to Cape Island from all over to see him."

"Well, there's no doubt Harry of the West is a great man." concurred Georgina. "He has run for president I don't know how many times. And, of course, even I have heard of the Missouri Compromise. The country would be at war with itself if it weren't for Mr. Clay."

It was with no little admiration that Harrah saw how skillfully Georgina was able to manage things. She had kept her husband at bay and the conversation flowing—all at the same time. His estimation of her went up considerably.

"I don't like compromises," gasped Brookens suddenly. "Whether it's slavery or the tariff. It should be one way or the other. And let the devil take the hindmost!"

Georgina looked across at Harrah, as much surprised by the outburst as Harrah was.

"Is there a chance we can see him?" she asked.

"Mr. Clay? Probably. He'll be staying at the Mansion House. And if he goes into the water at all, he's sure to find company."

"I would love to meet him," said Georgina. "I'll bring a pair of scissors to get a snip of his hair."

"He's still in mourning for his son," announced Brookens oarsely, without once turning his head from the ocean. "It would ardly do to go snipping for keepsakes."

"But it's the only chance I'll get to see him!"

"I'll see what I can do," said Harrah, feeling pity for Brookens relief that Georgina, not he, had to look after the man.

LADY IN A CARRIAGE

Harrah found himself thinking about Georgina more than he liked. He had not looked for Georgina to appear on Cape Island. He had relegated her to what he considered a former life, his years in the Pines, along with Bill Hostetler, Jesse Richards, and all the other phantoms of Burnt Tavern Road. Having discovered she was just across the street at Columbia House, he found her too close to home. Trescott and his children were a kind of special preserve—except where his house guests were concerned—and he wanted nothing in his past life to intrude on this.

Yet there could be no denying Georgina once she made her presence felt. She had a claim on Harrah, a claim staked by past intimacy. And he was not such an ingrate as to pretend it never happened. At the same time, despite her putting on a pound or two, he was still drawn to her. What attracted Harrah to the woman, apart from her considerable charms, was the prospect of immediate reward, instant sexual gratification. In the past he would no sooner meet with Georgina than the meeting would culminate in another meeting, this time in bed. And the idea kept Harrah in a constant state of expectation, like the prospect of dessert without having to wade through dinner. Yet no nagging guilt feelings troubled him. No worry about tomorrow interfered with the pleasure domes of today.

And so when a messenger told him that Georgina's carriage was waiting outside, Harrah left word with Rachel in the kitchen that he was stepping out.

Georgina Brookens sat in her hired carriage like the grandest lady of Columbia House, a parasol shading her from the darting sun of a fitfully cloudy day.

"I want to talk to you," she said. "Do you have the time?"

Harrah nodded.

"You can drive if you wish. I'll send Mr. Talbot to lunch." So saying, she waved the carriage owner from his seat.

Harrah helped Georgina step forward, then he climbed up beside her and took the horse's reins.

"Drive along the strand," said Georgina with more authority than he had ever seen in her. "The waves are wonderfully aggressive today. I love watching them splash against the beach."

Harrah did as he was bid and their horse was soon trotting briskly on the hard white sand, heading north away from the guest houses. The clouds overhead looked majestic in their puffiness, and in their gray-white underbellies Harrah saw the promise of late afternoon encounters.

Georgina put away her parasol, which was being caught in the breeze, and adjusted her gown which had become twisted at the waist.

"You keep avoiding the subject, so I'll come to the point. I know you want me. Nathaniel. And I want you, too," she said, not without a flush of cheek. "But I'll not give myself to you until we talk of marriage."

"There were no preconditions before," thought Harrah, and at last he said as much.

"In those days, at least one of us was not available."

"You're not available now," Harrah reminded her.

"I know that. But I will be soon. It's only a matter of time. Very little time for George, I'm afraid." She saw the change in his face. "Such talk offends you? I'm sorry, but it must be said. I can't go on waiting for you forever."

Harrah said nothing in reply. He merely chucked the horse into a more rapid trot.

"It may be that you never thought of marriage," Georgina continued, bravely riding the bumps. "But I've been thinking about it a great deal of late. I don't like being alone. No woman does. At the same time I want a husband I can be proud of. And, of course, I've always been partial to you."

She squeezed his arm and brought it in contact with her breast. "And now I'm in a position to do you some good. Or will be."

"I don't want your money, Georgina."

"It will be our money. One doesn't have to look far to see that you're not progressing. Your daughter will need a dowry some day. And your son a good business or profession."

"I can take care of my children."

"Yes, and you can take care of me, too. With George I had to subdue my feelings. Put them in wraps, so to speak. I was never really attracted to him. With you—"

"You needn't flatter me, Georgina."

"It's not flattery. It's God's truth! I'm at my peak now as a woman. I don't want to squander my passion on a man I don't care for. But I've always cared for you, Nathaniel. I don't have to tell you how much!"

Harrah drove to the far end of the strand as Georgina nodded to some acquaintances in a passing carriage. Satisfied they were alone, though the sky hung low above them and gulls careened in the gray air, Harrah put his reins down.

"I have to speak honestly. Georgina. I don't like talking this way with your husband at death's door."

Georgina curled the corner of her mouth.

"If we're being honest, why don't you say that it would be no different if George were already dead."

"I don't know how it would be," said Harrah, turning away to look at the ocean. "I know only how it is now."

"Tell me that you're not attracted to me."

"I can't deny it. I am."

"Tell me you wouldn't go to bed with me if I invited you."

Harrah told her nothing.

"See. I know I still have power over you."

"It's not power, Georgina. It goes by another name."

"No matter. It's there. I can see it in your eyes. I can feel it in the warmth of your fingers," she said as she lay her hand on his.

"I think we'd better go back now," said Harrah. "I have a business appointment at noon. Had I known you were coming—"

"Will I see you tonight. They have an especially good supper at the hotel."

"Not tonight, Georgina."

"Tomorrow, then."

"Tomorrow," said Harrah.

Picking up the reins, he started back. As it started suddenly to shower, he did not have to apologize for racing the horse.

SOMEONE ELSE'S WIFE

Harrah had gone to Schellenger's Landing the next day for some lumber. It occurred to him that a portable boathouse or two might be just the thing for ladies who like to go bathing but were too modest to do so in public. Congress Hall had these boathouses, and so did the New Atlantic and Columbia House. But as far as Harrah knew, none of the smaller guest houses could boast of such an amenity.

At the lumber yard black-eyed Jeremiah Schellenger suggested Harrah hire a carpenter to make them, not waste his time trying to build them himself. In this way Harrah would have the boathouses in place by the end of the season. Joseph Ware was available just now. Did Harrah know him? He lived on Lafayette Street, near Franklin, and was experienced in these things. Harrah returned home without the lumber.

"A lady was here to see you," said Willacassa, greeting him at the door. "She told me her name but I've forgotten it. A Mrs. Brook or something."

"Mrs. Brookens?"

"That's it."

"Did she leave a message?"

"Only that she would stop by again after noon." Willacassa looked crestfallen. "I don't like her, Daddy."

Harrah was set back. "Why not, Willacassa?"

"She's too frivolly. She's all washed and perfumed, not like Rachel."

"But Rachel is a cook. And Rachel is a good deal older than Mrs. Brookens. Besides, I thought her perfume had a pleasant fragrance."

"It does," admitted Willacassa. "That's why I don't like her."

Harrah could see that Willacassa was not rational on this score, even allowing for her age. He tried to glide past the subject.

"Is Jonathan up yet?"

"No he's still in bed. He's complaining of a sore shoulder. I don't think he wants to sweep the porch today."

"Maybe his shoulder is sore."

"It wouldn't be sore if he could sneak in at the Kursaal."

"He's too young to go there."

"But he sneaks in for the entertainment and they let him stay."

"I don't know why she doesn't like me," said Georgina when she called again at noon. "But it's clear she doesn't."

20

"Maybe she finds you too pretty."

"Never mind how she finds me. Do I please you?"

"Always."

"Then why don't you act it?"

"I thought I did."

"By being more demonstrative, I mean. I'm practically throwing myself at you. And all you can do is say polite nothings."

"I thought I had shown you Cape Island."

"Oh, you've been dutiful. We've been to the Kursaal. And it is magnificent. And I've seen not only the Virginia Serenaders but that Italian opera burlesque, *Stuffo*—"

"And you've been to the Atlantic Hotel's grand subscription ball."

"But you didn't dance."

"I don't like dancing."

"I don't either," she confessed. "I'd much rather—"

"Let it remain unsaid," cautioned Harrah.

"Why?"

"Because I prefer it that way."

"But I want to be close to you."

"You're quite close now," he noted as she pressed her body against his.

She pulled away, the nearest thing to anger in her eyes, Georgina who was almost never angry.

"Tell me you don't like it," she said.

"Not here. Not now."

"Your house is half empty. You said no yourself. I am sure we can find a room there and some privacy."

"My children live in the house. And my cook. It would be quite impossible. Besides—"

"You're worried about George?"

"Not worried. It doesn't seem proper somehow."

"You didn't worry about him before."

"He wasn't sick before. And that was some time ago. And he wasn't your husband then. I like to think we've changed our ways over the years."

"Some things never change," whispered Georgina, pressing closer to him. "Certainly not my feelings for you."

Nor did Harrah's feelings for Georgina change. He was as attracted to her as ever, though it was not just her face that attracted him, beautiful as it was. Georgina had a quality of carriage, both as a woman fully dressed and—as he imagined and remembered her—in the nude, that intrigued him. It was not simply the way her flesh clung to her figure that fired his imagination, but the allure that time and place lent her body—its fullness, its suppleness, its proportions exact to his every expectation. There was something of the eternal woman about Georgina that could not be denied.

And yet as the day of George's death drew nearer, he felt himself choking for breath.

Either Georgina did not notice this or she was too preoccupied with her own situation to see Harrah's growing discomfort. She began visiting him in the evenings.

"You have no idea how much money he has made, Nathaniel. Despite all the accidents and the fires and the cost of constructing new lines, the railroads are a gold mine. And this money will be mine soon." Georgina was unable to hide the sense of comfort her prospects seemed to promise.

"He's not dead yet."

"You've seen him. How long can he last?"

"I wish him well."

"And so have I these past three years. I don't want George to die—any more than I want to die. But he will die—and soon. And there is no use pretending he won't. Besides, I've been a good wife to him. You can see. I take good care of him. I always have."

"I know that, Georgina."

"Then why shouldn't we plan ahead?"

"Because there's no planning ahead in this instance."

"You don't love me, do you? I suppose you never loved me, not even in Elfreth's Alley. But you were fond of me. I almost think you like me better when I belong to someone else."

That was true, thought Harrah. He need not concern himself when she was someone else's wife. The husband was there to look after her. And he was free to think of other things.

"Remember this?" she asked.

Georgina pulled out a miniature of a portrait he had done of her some fifteen or sixteen years ago on Burnt Tavern Road.

"How did you get that?"

Harrah was as impressed with the rendering on ivory as he was with his own portrait whose quality had somehow been recaptured in miniature.

"I had an artist copy it in Philadelphia. A good likeness, don't you think?"

"Yes, a creditable job." Harrah carefully inspected the work of art, looking for differences between what he remembered of his own work and the copy, but finding mostly similarities. And for a moment he was brought back to the time when he had done the sketch. How beautiful was Georgina then! How pure and innocent! How much an ideal of young womanhood!

He looked at her to see where the years had taken their toll. But in all fairness he had to admit that Georgina had changed only a little. Fuller, more mature, a trifle paler. But for an instant as he stood looking at her Harrah could see Georgina exactly as she had been. And the reincarnation had a magical effect on him.

Yes, this was the Georgina of the miniature. The Georgina of the coach stop on Burnt Tavern Road. The Georgina of Elfreth's Alley and Quaker Bridge. The Georgina Cassandra had been jealous of, the dark side of Cassandra's fair coin.

Georgina leaned forward, tilting her handsome head.

"Do you still think I'm beautiful?"

She was not unaware that in the leaning her breasts swelled and offered a shiny prospect. Still Harrah made no move to oblige her.

Dismissing his inaction as timidity, she took his hand and guided it along her cheekbones, first one side, then the other. Delicious to the touch they were. Then she guided his hand to where his eyes were fixed. It glided over the smooth, heaving curve of her breast until it came in contact with the clove tip of her nipple.

"Come now, was that so bad?" she asked.

Without waiting for an answer, she undid her gown and made room for him to lie down beside her.

NIGHTWATCH

Willacassa pulled her bed next to the window of her bedroom. There was a mark on the wall where the bed rested against it before, and Willacassa felt compelled, as much out of ritual as out of neatness, to line the bed up against the mark. Leaning on her arms, she peered out. At bedtime she loved to watch the coming and going of the fashionable guests across the street at the Columbia House. Not only was the Columbia House a grand structure, dwarfing everything in the area, it was the watering place of grand people. There were carriages aplenty, segars aplenty, grand gowns, and white and black suits. Servants perpetually hovered about. Bright lights shown from inside the hotel. Lamps were lit outside. And there was music almost every evening.

But most important was the social coming together of men and women. Though not all of them were young—venerable beards and gray hair were dispersed among the guests—all were enjoying the free banter and easy camaraderie of Cape Island. And there were the usual flirtations.

Willacassa understood the byplay and the sexual divertissement of the July and August visitors. She had seen it all her life. But the variations of what she had come to know as Cape Island romances never ceased to intrigue her.

One young woman came home to Columbia House every evening with a different caller. How did she manage it? Willacassa wondered. Though pretty, she was no prettier than the other girls. Yet the men hung about her as though she were royalty. And why couldn't she find one man to suit her instead of shooting down with nervous laughter a whole covey of young swains, any one of whom could capture a sensible girl's heart.

Another woman teased her escort mercilessly. Willacassa could clearly hear her voice on the almost deserted street.

"You foolish boy! Of course I love my husband! Does he mind my seeing you? Of course not! He's a gentleman, and gentlemen show no jealousy. Besides, he knows your father. They went to the University of Virginia together. I can't help it if I was plucked from innocence before I was sixteen. He was a man of the world! I let you kiss me because I felt sorry for you. And I thought you knew something about dancing. Of course I'll have lunch with you tomorrow. But you must quickly mend your ways. I will tolerate youthful indiscretions. But I will not be embarrassed by protestations of sincerity."

The young man was so grateful for small favors that he all but stammered a good night. And then he managed to trip himself as he stumbled back to his waiting carriage.

And then there was the young couple who were obviously madly in love and hated to part. They invented at least a dozen ways to put off saying good night. Willacassa recognized the perennial purity of their relationship but knew almost instinctively something would get in their way. After all, there was always a stumbling block, wasn't there? An acquaintance had told her father that money, sectionalism, or religion invariably provided the obstacle to lovers marrying. But her father shook his head and said, "No, it's the perversity of the race. We're not good at matchmaking. When there's a good, a lasting marriage, it is more by accident than design."

And Willacassa believed what her father said—at least for other people, if not for herself. For one day she would fall in love and no perversity on earth would prevent her from marrying the young man of her choice!

She soon grew tired. The carriages drew up only infrequently now. The beach strollers had long since returned to their rooms. Window lamps dimmed. Servants were seen scurrying about, collecting chairs and tidying things up. And the roar of the ocean had found its night voice.

She had almost turned away from the window when a figure emerged from under the porch of Trescott House. It was a woman's figure, and she hurried across the street towards Columbia House. When she reached the building, she climbed the stairs to the verandah, then was swallowed up in the shadows of the wrap-around.

The next morning, when she was not so tired, it occurred to Willacassa who the woman might be. She was beside herself with jealousy. At first she refused to believe that it was her father the woman had visited at such a late hour. But Willacassa had previously seen Mrs. Brookens in his office. And though she had not thought so at the time, she now realized that he had known her at an earlier period in his life, in a place other than Cape Island. There was something insinuatingly possessive about the woman's attitude toward him. Mere acquaintanceship could not explain it. And this disturbed the girl.

Naturally Willacassa did not think of her father as being unattractive to women. In fact, she much preferred to think that all the eligible ladies in town wanted to marry him. By eligible, of course, she meant young ladies (but not too young) who were beautiful, modest, and reasonably wealthy. She had only one reservation. She reserved the right to pick her mother's successor, or at least give her the stamp of approval.

But this woman, the woman who emerged from Trescott House the night before, Willacassa dismissed out of hand. And the girl was determined to find out more about her so that she could like her less than she did already.

IN THE LIGHT OF DAY

Harrah lay in his bed long after his usual hour for rising. His head was full of images of the night before, and he could not begin a new day until he put them to rest. It had been a long time since he had known a woman. It was more than five years ago that he had lost Cassandra, and at no time since had he taken a woman to bed. He had almost forgotten the deep satisfaction of such an act. He had thought of himself as puritanical but it might just as well have been priest-like. Though he had taken no vow of chastity, Harrah had for five years been as abstinent as a Capuchin monk.

But Georgina had changed all that. In one night she had awakened in him all the passion that had lain dormant for so long. She had called upon the physical responses common to all men and he had feasted upon every moment of their time together. What strangeness he might have experienced with another woman, a woman he had not known intimately before, he did not feel with Georgina. For he had done with her before—in Elfreth's Alley many years ago and then at Quaker Bridge in the Pines.

He knew what to expect from Georgina. And she did not disappoint him. When he was through making love to her, he felt in no way deprived or cheated or denied. She had given herself fully, and he had accepted her gift.

More than that, he knew that she had a deep affection for him. By turns she whispered sweet notions into his ear or hummed or even shed a tear or two as she rested on his chest. And where only the ends of her hair had been windblown before, all of her hair now swept out of control and caressed him as she fondled him with her fingers and her lips.

How, after an hour or so, she was able to pull herself together and get into her gown once more and pull all her loose strands together into an acceptable mode, Harrah did not know. But manage it she did. And in a moment she was out the door, out of his house, and gone.

It was only later, when her sensuous presence was no longer in the room, that Harrah could even begin to think about his other feelings for her.

The truth was, Georgina never matched the ideal image he had so skillfully fashioned when he did that superlative charcoal portrait of her at a coach stop on Burnt Tavern Road. Actually, no woman of flesh and blood could. And when at Elfreth's Alley he learned more about her life and those frailties and weaknesses of personality that seemed to flaw the character of an otherwise generous and gracious young woman, he could not bridge the gap between portrait and person. There was an inevitable division here that could not be reconciled. And while he preferred Georgina in the flesh, he preferred the portrait to the Georgina he had come to know.

And then, like him or not, there was George Brookens. Of course, in no physical sense was Georgina his wife anymore. George was now beyond the pale of the living, hanging on with what lung was left him. It was cruel to think this way, Harrah admitted. But it was life that was cruel, not any thoughts Harrah might have on the subject. If Georgina was not free to marry, she was at least free to seek solace. And if she had her way, Harrah would provide that solace.

Willacassa brought him some coffee and fresh biscuits and set them down on his bed table.

"Remember Mrs. Brookens?" she asked.

"Why, yes," said Harrah, sitting up to fill his cup.

"Did you know her long?"

"I once did her portrait in the Pines. Many years ago."

This was unexpected news to Willacassa. Her mother was supposed to have been the artist in the family. She knew her father had a talent for drawing, but she never thought of him as a portrait painter.

"Is she a nice lady?"

"Very nice."

"That's strange."

"Strange? Why?"

"They don't say nice things about her at the Columbia House. They say she's lolling around while her husband is dying."

Harrah pushed the coffee aside and reached for his robe. "Is that what they say? That's most unkind of them."

"Then you don't agree?"

"I agree that her husband is dying. But I don't think Mrs. Brookens must go into mourning just yet."

"They say she has a boyfriend."

"They do?" Harrah rose to cut short the discussion. "That's no one's business but hers, Willacassa."

"Are you the boyfriend, Daddy?" Willacassa regretted asking this the moment the words were out.

"That, my girl, is none of your business." Harrah's fatherly beneficence evaporated into cold fire. "In fact, the details of Mrs. Brooken's life are none of your business. A child should concern herself with children's things and not try to understand matters that are beyond her grasp.''

"I'm not a child!''

"Don't contradict me!''

It was the not the first time Harrah had been angered by Willacassa, but it was the first time he found his anger so pointed with annoyance. He did not like having to answer to his own daughter. A not-quite-ten-year-old child at that! He did not like it at all!

Yet no sooner did Harrah shout at the girl and see the love fade from her eyes than he despised himself for having lost his self-control.

"I'm sorry I became angry,'' he said after a pause. "But I do want you to understand that there are limits to the questions you can ask.''

"Is that an apology?'' Willacassa asked.

Harrah was not about to admit he had gone too far. It was too early in their relationship to concede his fallibility. Nor did he want to encourage any further investigation by Willacassa.

"No, not an apology. An explanation.''

"Well,'' said his daughter, "an apology wouldn't have been accepted anyway.''

ARMAND FOUCAULT

Armand Foucault—Harrah did not know whether to call him Mister or Monsieur—came from New Orleans by way of Mississippi. For the past two seasons he had taken an upstairs room at Trescott House with no view of the ocean. Even when the newly built Columbia House drained off some of Harrah's clientele, Foucault stayed on, never complaining, never demanding very much in the way of service. He would leave Trescott House after the dinner hour and for the rest of the day and evening find planters and railroadmen to play cards with on the breezy verandahs of the New Atlantic or Columbia House. And every night when he would return, Foucault, as a token of his winnings, would leave a bottle of wine on the dining room table, a bag of hard candies for Jonathan, and a sparkling trinket for Willacassa.

"Don't you ever lose?" asked Harrah one warm night when, sitting on the porch beyond his usual hour, he met Foucault coming home.

"But never." This was not a boast on Foucault's part, only a statement of fact.

"How do you do it?"

Foucault pulled up a rocker and sat down. "I am a gambler, Mr. Harrah. It is my profession. These gentlemen I play with, they are at best amateurs. When amateurs and professionals meet, there can be no contest."

"But don't the losers complain?"

"Why should they? They have money enough. And I don't play for high stakes. Just enough to live on. To live comfortably, of course. The only ones who complain are the hotel proprietors. They complain for two reasons. First, because I am not a guest there. Second, because they frown down on gambling. One would think they were Quakers instead of entrepreneurs."

"Why don't you stay at the New Atlantic or Columbia House? Then at least the first objection would be removed."

"Because I would no longer be a complete stranger. Men prefer to lose at cards to strangers than to friends or even passing acquaintances. Besides, if I stayed at these hotels it would appear as if card playing was sanctioned there."

"Perhaps you can teach me a trick or two, Monsieur Foucault. Then maybe I could afford to keep Trescott House going."

"You are experiencing financial difficulty?"

Harrah nodded.

"I had some idea this was so. Even with Mr. Henry Clay on

the Island, you have not yet filled your house."

"Exactly."

"Yet you keep a good establishment," observed Foucault.

"Thank you."

Armand Foucault twisted his moustache, then pulled on his long, thin nose.

"I have a suggestion to make—if I may be so bold."

"Please do."

"You are a fine man, Mr. Harrah. A man of principles and scruples. Forgive me for saying so."

"There's nothing to forgive. I would think."

"You offer a good room at a fair price. And you believe that will bring you patrons."

Harrah nodded.

"Not in these times." pronounced Foucault. "The guest houses on Cape Island are getting bigger and better. Every year a new one makes its appearance."

"So Trescott House must get bigger and better, too."

"Either that. Or—"

"Or what?"

"Become more expensive. No that is not the proper word. More exclusive would be better."

"I don't follow you."

"You should cater to pride, Mr. Harrah, the pride that comes with self-importance or, better still, self-indulgence. There are those who are inordinately proud of their station in life. You should charge them twice as much as the Columbia House."

"Twice?"

"Maybe three times. Otherwise they feel cheated. But of course, you must provide nothing but the finest. The finest food, the finest wines, the most intimate tables. And—"

"I'm afraid you're already beyond me."

"And you should not only permit gambling, but make of it a major attraction."

"But I'm no gambler."

"You misunderstand me, Mr. Harrah. Trescott House itself should not indulge in gambling. Put that out of your mind. It should merely *permit* it. If wealthy gentlemen wish to indulge in cards, let them. Why shouldn't they have a comfortable place to play while they are taking advantage of the refreshing waters and the sea breezes of Cape Island? And why should they not have the finest wines to drink while they are playing? Not your local vintages. But Spanish and French wines. And brandies and cognacs!"

"You're serious?"

"I'm not only serious. I'll lend you the money should you need it. And I will ask only the principal in return."

Harrah leaned back in his chair and sighed deeply. For himself money was no problem. He could live on nothing a year. At least, next to nothing. But always he came back to the one drawback, the one sticking point. His children required sustenance. Not the luxuries of life, yet something more than the bare necessities. It was as if he had to compensate them for the absence of a mother. And in Foucault's plan he saw a chance to turn his fortunes around.

"But why, Monsieur Foucault? Why would you do this?"

"Let us say, I am tired of going across the street to pitch my tent. I'm not a nomad. I'm a Frenchman. As a guest in *your* house, I will not be frowned down upon. That much I know."

Harrah did not contradict him, but he did proceed with caution.

"And you will place a limit on your winnings—just as you do now?"

"I will even manage to lose on occasion." He took Harrah's arm. "And if ever you want me to leave Trescott House, for whatever reason, I will leave. Is that generous enough?"

"More than generous."

"Then you will take up my suggestion?"

"Let me think about it, Monsieur Foucault."

"But of course, think about it. Think about it all winter if you must. But come spring I hope to see a much altered Trescott House, my friend."

A MESSAGE

Harrah awoke in early morning with the sense something was wrong. He pulled on a shirt and trousers and, opening his door, walked the hallway into the reception room. He found a colored youth standing there. Harrah recognized him as one of the servants at the Columbia House.

"What is it?" he asked.

"Mrs. Brookens wants you. Right away, she said. Everyone was asleep here and I didn't know where to find you. So I jus' waited."

"Thank you. I'll be there in a few minutes."

When he met Georgina at her verandah door she was as ashen as sailcloth.

"He's dead." she whispered. "George is dead. Come inside."

Harrah expected to find the man lying in his bed but saw no one else in the room.

"They've already taken him away," said Georgina, trembling with agitation. "They weren't too pleased that he died here. Don't want the guests to know."

Management was the same all over, thought Harrah.

"I thought he was too quiet. No coughing and no complaining. Still it was a shock to find him dead."

"Where did they take him?" asked Harrah.

"They're shipping the body to Philadelphia. Put him in a coffin straightway. He'll be buried beside his brother, Gerald."

She narrated all this as though the hotel people had made the plans, not she.

And Harrah realized why Georgina was so determined to settle things between them beforehand. Until George died, she could put up a bold front. With his death, she was unable to think clearly, to act decisively. She seemed to have drifted into a passivity of fear and paralysis.

"I must go with him, I suppose. There's no one else to look after the body."

"I'll accompany you, Georgina."

"No, that won't do at all. Your place is here—with the children. Even I can see that."

"Rachel will look after them."

"And there's Trescott House to look after. It's the height of the season."

"But—"

"I'll be all right, Nathaniel." She stirred herself into motion and began to gather her belongings. "There is one thing you can do. Collect all my baggage and George's things and send them to the Landing. I can't stay in this room another minute. I've got a carriage waiting."

She turned to look at him.

"So you see. That's how things are, Nathaniel." She even managed a smile. "You have my address in Philadelphia. When the summer season is over, come to see me. But only if you're ready for marriage. If you don't come, I'll understand."

She moved toward him and embraced him, kissing his cheek. Then, pulling away, she looked over the verandah rail to see if her carriage was still waiting, and started downstairs.

1848

YEAR OF CHANGE

OUT OF THE PAST

Harrah pulled open his door to find the figure of Captain Irons framed in the doorway. It was all unexpected, totally without warning. With his dark seaman's cap, a seaman's coat, and a seaman's bag on his shoulder, he looked like a blue derelict floating on a gray sea of mist. It was almost two years to the day of the wreck of the schooner, *Carolina*. And the apparition recalled the day with funereal precision.

"You seem surprised to see me," muttered the captain in a hoarse, salty voice. "And not at all receptive."

"I am surprised," Harrah managed to reply.

"I would have been back sooner," announced the captain. "But it took a while to outfit a new ship. And then I had contracts to fulfill. In fact, I've returned from the Orient and the subcontinent. Where is the lovely lass?" He set down his bag which Harrah could see was laden with gifts.

"Willacassa?" Harrah maintained his grip on the doorknob. "I don't think she'll be pleased to see you."

Captain Irons made a move to enter the house but Harrah blocked his way.

This startled the captain. A cloud dimmed his teal blue eyes. "Why not?"

"I think you know why. Some of your cargo washed up on the beach—a few days after you sailed on the *Palmetto*."

"Cargo?"

Harrah dismissed his attempt at bewilderment. "I believe you know what I mean."

Captain Irons let the smile fade from his lips. "I don't know what you're talking about, Mr. Harrah. What's more, I don't like it. I do hope you haven't poisoned the child against me."

"It wasn't I poisoned her. But see for yourself." Harrah gestured in the direction of Ocean Street where the girlish figure of Willacassa was approaching. "I suggest, Captain, you give her time to get over the shock of seeing you."

Captain Irons stepped back. He quickly removed some gifts from the seaman's bag—a carved figurehead, Chinese slippers, and a silk robe—and made ready to greet the girl. A broad smile spread like a canvas across his face.

Upon seeing him, Willacassa stopped. No ghost could have frightened her more, no gargoyle prompted a more harrowing response. She took a faltering step backwards.

The captain called to her, still beckoning with his smile. "Don't

you know me?'' When Willacassa did not reply, he ventured, ''I'm Captain Irons, my angel of mercy. Surely you remember. Look what I brought for you!''

Willacassa was frozen into silence.

''It can't be that you've forgotten me already''

Willacassa paled to a bleached muslin. ''I haven't forgotten you. And I haven't forgotten the ship *Carolina* either.'' She paused and summoned up her courage as Captain Irons drew a step closer. ''And I haven't forgotten the slaves who washed up on Higbee's Beach. Your slaves.''

''Slaves?''

''Yes, chained together—and dead.''

''Who put such nonsense into your head?'' The captain's mood slipped into scowling dudgeon.

''It's no nonsense. I saw them.''

''But I had nothing to do with that, my child.''

Willacassa fixed a disbelieving eye on him. ''Reverend Williamson says it's not right to lie—even when the truth is painful.''

''But—''

''I think this has gone far enough,'' indicated Harrah. ''The girl is terrified of you, Captain. Can't you see that? You'd better go now.''

Captain Irons turned towards Harrah and spoke as though Willacassa were not within earshot.

''It was only cargo, Mr. Harrah. You're making a great deal of fuss over nothing. The poor devils are better off now than they would have been sold on the block.''

''That doesn't alter your role in all this.''

''My role? My role was honest broker. If it weren't for the storm, those blacks would still be alive. There's a flourishing trade in slaves. I didn't invent the institution—or the hypocrisy that goes with it.''

''No, but you've made a handsome profit from it. And would have made more if not for what you did.''

''You can't prove that I did anything.''

''No, I can't. But you know better than I what I mean.'' Harrah descended the steps of Trescott House and put a comforting arm round his daughter, who had not yet moved.

''There's nothing more to say, Captain. You'd best be on your way.''

Captain Irons turned a ghastly gray. It was clear he had no patience with disappointment, no tolerance for criticism. The reception he expected had turned to acid, even scalding indictment. And here was a man who enjoyed unbridled authority aboard ship. But the seaman was not one to be outdone or easily put off. His eyes shifted about looking for some advantage, for something to turn the tide his way. He even glanced over his shoulder at the towering structure of the Columbia House across the street. And then he seemed to hit upon something.

"Well, you may choose to take me to task, Mr. Harrah. But I haven't forgotten your past kindness." He dug deep into his pocket, then pressed a roll of bills into Harrah's startled hand. "Here this will get you started, something to help you run with the Columbia House cross the street. One can see at a glance that things aren't going too well with you."

At first Harrah did not comprehend what was taking place. Yet it was plain enough. The anger that at last surged up almost choked him. This time he made no attempt to hide his feelings. He rudely shoved the money back into the captain's hand.

"I don't need your money, Captain Irons! Nor do I want it—tainted as it is! Trescott House may be hurting, but we sleep at night with clean conscience!"

Captain Irons could see he was making no headway. If anything, his last thrust had all but wrecked him. Setting the gifts down on the porch steps (in the hope that in time Willacassa's curiosity would get the best of her), he slowly backed off. He buttoned his coat, cloaking himself in rectitude, in New England self-righteousness. The stain of shame that Harrah and shivering Willacassa saw did not exist for him. He slowly moved away, like a ship pushing off from a forbidden dock. Then with a seaman's roll he turned and without a look back disappeared behind a lumbering wagon and a wave of dust from Ocean Street.

RENOVATION

After mulling over it all winter and weighing the pros and cons, Harrah decided to take Armand Foucault's advice. Dipping into what savings he had left and arranging with the lumber yard for credit until August so as not to leave himself without a few dollars in his pocket, Harrah drew up plans for a major renovation of Trescott House. Working feverishly in order to begin work by spring and be finished by early summer, he put to use his limited artistic talent and sketched in the new rooms as he had envisaged them.

What he had in mind was simple enough: a drawing room, a parlor, a reception hall, and a small office—all on the main floor. The drawing room, of course, would be a euphemism for a game room with card tables, a liquor cabinet, and a lavish humidor set on a mantel of black Italian marble. All this would be fashioned out of a gutted downstairs where walls would give way and doors would be replaced. A new dining hall with wide windows would be added to the existing structure, and Rachel's kitchen enlarged so as to be big enough to feed Harrah, the children, and Rachel herself. Though the bedrooms would remain the same, their furnishings would be upgraded. Each room would have fourposters with canopies, Sheraton nightstands, dressing tables with beveled mirrors, and camelback loveseats.

And though there was all manner of reservation and hesitation on his part about the design of the new rooms, the fixtures and the materials used—so much so that he wanted to give it up and start all over—Harrah forced himself to push on. At last, closing his mind to further alterations, he showed his plans to local carpenters.

"I'm too busy, Mr. Harrah. Already signed up with Columbia House," said one.

"I've got some cottages to build on Hughes Street," explained another.

"You should have looked me up earlier. Can't help you now. Maybe Daniel Hand at Cape May Court House has some free time. He's an architect and builder. But he gets one dollar and fifty cents a day."

So Daniel Hand it was. And Jacob Hand. And a young carpenter new to the Island, by the name of Sawyer. Trouble was, Harrah had to put the Hands up for the night, and feed them. Rachel was beside herself.

"You tear up my kitchen. Then you want me to feed your

hands."

"Their names are Hand, you know."

"So long as they don't eat with their hands. It's bad enough feeding you and Jonathan and Willacassa. That Henry Sawyer, young as he is, can eat for two!"

"At least he doesn't sleep here," said Harrah. "And just think how elegant your kitchen will be when we're through."

"Elegant, maybe. But more to scrub, for sure!"

He knew how Rachel, uncomplaining as she normally was, hated dislocation. But the job, the enormous job of transforming Trescott House to a first class "cottage" was under way.

Nothing, however, that Harrah needed except for lumber and labor could be gotten locally. He had to send to Philadelphia for furniture, wallpaper, linens, dinnerware, glass, and all of his silver. Fortunately he had made friends with a number of the steamboat captains. When their meetingplace at Commercial House had become unavailable, Harrah had offered Trescott House to the pilots free of charge. In gratitude Captain Devoe carried new wing chairs back from Philadelphia on the *Wave*. Captain Davis, who ventured as far as Frenchtown, Maryland, purchased several sets of silverware at a "steal," but forgot to deliver them until the *Ohio* had been laid up at the Landing with one of its two boilers blown. And Captains John Payne of the *Mountaineer* and Wilmon Whilldin, Jr., of the sometime scheduled *Napolean* vied with one another in providing rolls of wallpaper designed in the latest modes. Harrah used the blue paper in the drawing room and the maroon in the dining hall.

Armand Foucault was singularly impressed when at the end of June he arrived at Cape Island and took a carriage to Trescott House.

"I cannot believe it! The change is miraculous!" he said, disembarking. "If you don't rent every room this season. I'll pay double my rate."

'You'll pay double anyway," laughed Harrah, noting his luxurious white suit. "I've quadrupled my price. But you will get in for half the going rate. After all, you're the godfather of this whole project."

"How full are you?"

"I've rented twenty of the twenty five rooms through July. And the visitors haven't even begun to arrive yet."

"But how?"

"Through the mail. This advertisement in the Philadelphia newspapers turned the trick."

Harrah pulled out a clipping and showed it to Foucault.

"Trescott House, Cape Island

Newly refurbished with elegant, comfortable bedrooms, reading and drawing rooms. The subscriber welcomes single and married gentlemen and their wives (no servants) for games, cards, the finest cuisine and wines, and all the benefits of the seashore.

Trescott House is not inexpensive, but given the quality of its accommodations its rates are considered acceptable.

A new and commodious stable and carriage house has been built on the premises.

N. Harrah, Proprietor"

"It's the word 'cards' did it," ventured Foucault. "You cannot underestimate the gambling instincts of the rich."

Harrah smiled. "My son thinks the new stable and carriage house turned the trick. But Willacassa reminds him that these people don't bring their own carriages. They hire out."

"And what do you think, Mr. Harrah?"

"I credit the words 'not inexpensive.' It's not the rich so much. It's those who would be seen as rich who want the privilege of spending more for what they get. And I will try to oblige them."

"Well, we'll learn soon enough," said Foucault, helping himself with one of his bags while Harrah helped him with the other. "I shall make myself known as soon as I wash up."

By noon he had struck up an acquaintanceship with two planters from North Carolina and was soon seated at a card table with them.

"Did you win?" he was asked later.

"Of course not, Mr. Harrah. It is much better to lose at first. Winning comes later."

WILLACASSA'S REVOLT

Though Armand Foucault was pleased with the changes that had taken place at Trescott House, Willacassa was not. In the eleventh year of her young life she found that the new guest policy was making demands on her that she had not been prepared for nor was ready to accept. She spent hours in the big restructured kitchen helping Rachel put together new dishes which were a combination of South Jersey sea specialties, flounder, bluefish, oysters, crabs, and clams, and Southern cooking, especially the breads, corn cakes, and desserts. It seemed Willacassa was forever stuffing things and mixing exotic sauces. And having sampled the wine that was tossed into the pots, she was not sure that the meals served at the house would pass inspection by a church-sponsored temperance society.

She also spent a great deal of time making beds and straightening up rooms even though her father had hired colored girls as chambermaids.

"You can't overwork them or they'll leave for one of the bigger houses," said Harrah.

"But what about me, Daddy. I want to go out and spend time with my friends."

"You will—later. What do you think your friends are doing at this time?"

Of course they were doing the same things that Willacassa found so tedious. Or else they were working on their parents' farms, those of them who lived in the surrounding countryside, pumping water, shelling corn, feeding chickens, milking cows, churning butter, and then carrying the cans of milk and tubs of butter to the springhouse. Or else they were in the farmhouse kitchen, kneading dough, baking pies, washing dishes, and carrying slops to the pigsty.

What Willacassa objected to even more than the drudgery of the kitchen and of bedmaking were the new limits placed on her by her father. No longer did she have the run of the house, stopping to chat or lavishly being complimented by her father's patrons. (She particularly liked the Southern gentlemen with their soft drawls and impeccable manners.) Now she had to use the outdoor backstaircase, which had been added to the house in the spring, when she wanted to go downstairs.

"When men are playing cards, they don't want to be distracted, Willacassa. As you grow older, I'll let you wait on them. Serve coffee or tea and cakes or liqueurs."

"I don't want to wait on them. I want to be able to sit in these big, comfortable chairs and look out the window when my chores are done. Especially on rainy days. I don't want to be constantly in my room. I want to see people and talk to them. And I don't want to have to go across the street to the Columbia House to do it."

Harrah understood. It was not his intention to deprive Willacassa of freedom of action. But he expected her to understand that during the summer months at least their house was very much a business. The activities she could engage in or enjoy during the off season had temporarily to be suspended.

All this had the effect of driving Willacassa outside. She would go to the beach for a swim, mixing uninvited with the young mothers and governesses who with their children had during their designated time come down from the Mansion House or Congress Hall to the bath houses. Though the ladies generally wore suits of tunics and pantalettes in matching colors, some with white collars, some with wristbands, Willacassa's only concession to conformity was a straw hat. She freely wore the red flannel tunic and white duck pants favored by the gentlemen when she went dashing and splashing into the turbulent water, meeting the breakers head on and resisting the undertow once she found her feet.

After she had dried herself out, she changed clothes and walked Cape Island from end to end. She noted the tenpin alleys, the tent-like shooting galleries, the new ice cream saloon, the taverns, and the young men on the beach playing town ball to the shrill shrieks of the spectators.

One of the young men she recognized as Henry Sawyer. But the discovery that he was among the ball players so startled Willacassa that she did not stay to watch. The husky young carpenter was one of the three hired by her father the past spring to work on the alterations and additions to Trescott House. She had watched him help separate the dining room from the rest of the house and convert the former dining room to a parlor and drawing room. He was by far the youngest of the three carpenters working on the project, and he became the object of Willacassa's first crush.

Henry Sawyer had come to Cape Island only that year. With all the regular carpenters engaged by the bigger houses, Harrah had to settle for the newcomer, sight unseen. But the young man from Lehigh County, Pennsylvania, proved to be very good with his hands. A happy-go-lucky, ruggedly built fellow, Henry was afraid neither of height nor of hard work. In fact, the more difficult the job, the more willing he was to undertake it.

Willacassa noticed all this and more. She noticed a good deal for a girl of eleven—too much, thought Harrah, for her own good. Where was the child he had known? The child he still wanted her to be?

Willacassa had never been interested in quoits. But as soon as the day's work was done and the sun was down, Young Sawyer ran down to the beach to play a game or two with other post-season workmen. Willacassa stood nearby to watch. She was surprised at how skillful he was with the flat rings. If Henry was not the best player in the group, he ranked very near the top. At least once every game, to burst of laughter and applause, he tossed a "ringer." Then just as regularly he would miss a sure shot. On other days, farther down the beach, he would hunt for horseshoe crabs, cleaning out the insides with a knife and hanging the shells to dry and "stink out" before he gave them away as wall hangings.

But most of all, for relaxation, young Sawyer liked to race his horse at full gallop down the strand. He had gotten hold of a sword which had seen action in the Mexican War, and he swung this over his head with all the aplomb of a United States cavalryman or a European Hussar. Back and forth he would ride, skirting the water's edge and kicking up a spray. He always ended his ride with a dip in the ocean, horse, rider, and all.

Willacassa was terribly impressed. So this is what young men did. Until now she thought all young men were grown-up boys, pranksters given to tormenting young girls and telling dirty stories. She had a crush on Henry Sawyer that pained her with its weight, and she followed him wherever he went. If Sawyer was not the most exciting person on Cape Island, she wanted to know who was.

Fortunately, young Sawyer did not take her seriously. "Go play with your dolls," he would say. "And stop bothering me. Banged my hand with the hammer instead of the nail—for all your talking. Next I'll be sawing my arm off!" He laughed so hard, he half frightened the girl.

"I'm only seven years younger than Henry. That isn't much of a difference," she confided to Rachel.

"Why, you won't be eleven until December, Willacassa. Find a boy your own age."

"My own age. My own age. My own age! I'm sick of such talk!"

"Then wait a few years. Four or five years, and try again."

"Who wants to wait!"

And then, as suddenly as it came, the infatuation wore off. Henry Sawyer began growing a moustache. And moustaches Willacassa could not tolerate. Her father wore no moustache. The handsomest men in town were clean-shaven. Why should Henry Sawyer cover his upper lip with wisps of hair? Barbaric! It was enough to declare a moratorium on men. Even when Sawyer shaved if off, Willacassa did not forgive him. He had been one person before, now he was another. Growing a moustache had altered his image for all time. Shaving it off did nothing to rehabilitate him. If anything, it showed a lack of decision. And Willacassa could love no man who was indecisive.

CARDPLAYERS

The presence of Armand Foucault at Trescott House was a mixed blessing. As a Frenchman and an adopted son of New Orleans, he stood apart from both the Southern and Northern gentlemen who sojourned at Harrah's establishment. His accent seemed at times more French than necessary. And his flamboyant dress—other men wore ties, but none were so elaborately knotted and stick-pinned as Foucault's—called attention to him at once. His ability at cards established him not only as a professional, but as a suspect professional. This did not, however, rule him out when the tables were set and fresh decks unwrapped, and three other men sat down "to risk a few dollars." If anything, his reputation as a cardplayer made him greatly sought after.

What disturbed his playing companions, one a judge from Maryland and the other two Carolinians, was not that Foucault won so often or that he won so much or even that he puffed voluminously on broad-leafed Havanas. This they could afford to overlook. What irked them was the freedom with which he aired his views.

"Actually, I think the Negro is a superior race. Which of us could work in the fields so long and endure the hot sun? Which of us could tolerate the lash and the knout? Which of us could laugh so good-naturedly in the face of adversity?"

"They're beasts, I tell you! Beasts of burden! They're no better than mules or horses or oxen. Once you think of them that way, you can be kind to them. Just as you're kind to your horse. But don't make the mistake of thinking they are human."

"I met a darkie once," said Foucault between puffs of blue smoke, "who could divine the weather and find water where there had been none before."

"Voodoo, that's all!"

"That may be. But it slaked a powerful thirst. Three of a kind, monsieurs. I do hope we will play again soon."

Foucault confided to Harrah that he made provocative statements to test his opponents. If they became incensed by what he said, he knew at once that they held poor cards. If they dismissed it with only slight annoyance, he knew their hands were strong and he'd best bow out. But Harrah suspected that Foucault said these things out of perversity. He liked to get his opponents' hackles up. A keen student of human behavior, Foucault took particular delight in riling people who considered themselves his betters. Better or not, he'd wager they were wrong. And he would be the better for it!

As if his taunts were not enough, Foucault tweaked his table companions with his views on the national and European scenes.

"What do you think of the nomination of Lewis Cass on the Democratic ticket, Monsieur Foucault?"

"I prefer his brother, Jack."

"I don't think that amusing. If not Mr. Cass, who do you like, Zachary Taylor?"

"General Taylor is a slaveholder. He was picked only because the whigs are divided between Clay and Webster. And because he is a war hero."

"Then you're a Barnburner, one of those anti-slavery Democrats."

"Exactly. Just as a farmer is willing to burn down his barn to get rid of rats, so I'm willing to burn down the Democratic Party to rid the country of slavery."

"And yet you're from the South."

"Yes, but in Louisiana it is the plantation owners who own most of the land and most of the slaves. In fact, more than half the population is Negro. There is little left for the rest of us—even though we're regarded as 'free men.'"

The shuffling and dealing of cards inhibited the smoldering reaction to Foucault's pronouncements.

"What we have in the States is not unlike what is happening in Europe. The February Revolution in France occurred because of the oppressive conditions of the working class. It doesn't matter that the Revolution was betrayed in May and June.The unrest has spread to Germany and Austria and Hungary. And one day it will spill over into Russia where the serfs are nothing but white slaves. So long as we have slavery, the oppression of one group by another, we shall have division and unrest."

"No," said Judge Forbisher. "So long as we have men like you, Monsieur Foucault, we shall have internal strife. The opposition to slavery comes not from the slaves themselves nor from your Northern industrialists who are only too happy to get their cotton at dirt-cheap prices. Nor does it come from Western farmers who are too busy working the land to worry about the blacks. No, it comes from stiff-necked Northern preachers and foreigners like yourself. Men who have no sense of race and no respect for social distinctions."

The two Carolinians concurred and sipped their wine as though drinking a toast.

"Be that as it may," announced Foucault. "I have two pairs, kings and knaves."

The judge lay down his cards.

"Knaves? I should have suspected as much."

By August, Foucault had won a small fortune from one of the Carolinians, a Mr. Ethelbert Timmons. The sum of money involved made Harrah uneasy. He did not want Trescott House to gain a reputation as an establishment that fleeced its clients. It was enough that the public knew gambling was permitted on the premises and that the management made no money from it, just as Columbia House made no money from the occasional card games played on its verandahs. Harrah feared that Foucault had gone back on their original agreement. By winning big at the card table, he was giving the impression that Trescott House was a fool's paradise. Harrah made up his mind to speak to the gambler but kept putting off the confrontation.

There had also developed a seething resentment of Foucault by the men who played with him. Whether the resentment was born of basic philosophical differences or of the heavy losses sustained or both seemed inconsequential at this time. The important thing was the sense that trouble was brewing. And Harrah did not know how to head it off.

"I don't see how you could have possibly won that hand," protested Timmons.

"But I did win," said Foucault matter-of-factly, pocketing his winnings.

"Must have employed voodoo," suggested the Southerner's companion.

"Probably has Negro blood in him," added Timmons.

"He's swarthy at that."

"Most Latins are," continued Timmons. "Tell us the truth, Monsieur Foucault. What are you? A quadroon or an octoroon?"

"Neither. Just a man with a skill at cards."

"I've insulted you, Mr. Foucault. Haven't you honor enough to demand an apology?"

"I heard no insult."

Foucault rose and prepared to go.

"Then maybe this will stir you," said Timmons, rising too and raising his voice so all could hear. "You're a cheat, Monsieur Foucault. You pulled that last card out of your pocket. In fact, I suspect you've been cheating all along!"

Armand Foucault had the professional man's thick skin, and honor or a suggested lack of it did not penetrate its surface. But the word "cheat" was a firebrand that seared even his tough hide.

"You will withdraw that remark," he said, stiffening with anger.

"I will not!"

"You will withdraw that remark or I'll stove your head in!"

" Stove my head in, will you? Monsieur Foucault, I demand satisfaction!"

Foucault's neck became scarlet. His eyes turned rheumy. He found difficulty forming his words. "A ducl, monsicur?"

"Yes, a duel."

"With pistols?"

"With pistols!"

"I accept. Just tell me where and when."

At this point Harrah, who had just entered the room, interceded.

"We don't permit dueling here, gentlemen. Nowhere on the island, in fact."

"Please, Mr. Harrah. There must be some place," insisted Foucault.

"Yes, Mr. Harrah. Our honor is at stake. The honor of your house."

Harrah saw that there was no assuaging hurt feelings.

"All right," he said. "Higbee's Beach."

"Fine!"

"But only when I tell you. I cannot risk losing my license. Is that agreed?"

Both men nodded. "Agreed!"

THE DUEL

Three days later, before dawn, Harrah knocked softly on the door of Ethelbert Timmons and awakened him.

"Today's the day." he said quietly.

Timmons, half asleep, his hair touseled and his robe twisted out of shape, nodded distastefully and muttered, "I'll be ready."

When Foucault opened his door, he was clearly nervous and irritable. "But why today? It's terribly windy out."

"That's just it, Armand," whispered Harrah. "My son tells me that Timmons is an excellent marksman. He's seen him at the shooting galleries and he almost never misses. With a strong wind whipping around and sand blowing in his face, he'll have trouble seeing at Higbee's Beach. What's more he'll have trouble keeping his bullets on course."

"Ah," said Foucault, looking noticeably better. "Then I may yet survive the day."

Harrah had hired a dearborn and held it in readiness for the occasion. When Foucault, Timmons, and his fellow Carolinian climbed into the carriage, Harrah pulled the curtains down. Then slowly, like a tumbril rumbling to the guillotine, the four-wheeled vehicle crept through the dimly lit, hushed streets and out of Cape Island.

It was tacitly understood that Harrah would act as Foucault's second. There was no one else to assume the role. And it was left to Harrah to count off the steps and give the order to fire once the duel was under way. Given the gravity of the situation, none of the men in the carriage, which Harrah drove, was inclined to break the silence that had stifled speech at the very start and that hung over the carriage like a shroud.

When the dearborn reached the approach to Higbee's Beach, Harrah tethered the horse and led the men through the sandy path that wound its way to the shore. The going was difficult, not only because of the deep, damp sand which swallowed the foot that plunged into it but because the wind was playing havoc with their hats and the wild, twisted foliage that surrounded them.

"I can't think of a more forsaken spot!" said Timmons, breaking the silence at last.

"Exactly," replied Harrah. "That's why I selected it. No one is likely to observe us here."

Timmons's second, Mr. Woodbourne, clung tightly to the box of pistols he carried under his arm. He concentrated solely on negotiating the footpath. When he did look up, he saw dark, fast-moving clouds racing at him. And so, pulling his hat down, he lowered his head once more and fixed his gaze on the shifting sand, much of which had crept into his shoes.

At last the foliage gave way to an opening and the men found themselves on a broad expanse of white beach. But a restless ocean, swollen with turbulence, churned only a small distance from them, its high, chopping waves rolling up the sand almost to where they stood before receding again. Timmons looked at Harrah in a gesture of protest for the unusual dueling conditions. But Harrah acted as though they were in no way out of the ordinary despite the fact he had never seen a duel on Cape Island and had heard of only one.

"Would you care to examine the pistols, Monsieur Foucault?" asked Woodbourne finally.

Even with the weather conditions as rough as they were, Foucault was not too keen on proceeding. But Woodbourne opened his oak case and in red velvet displayed a brace of handsome Parker pistols with spur trigger guards, good quality octagonal barrels, and swivel ramrods.

"They are not French or Belgian pistols," he said critically. "But I suppose they'll do."

"Take your pick," invited Woodbourne.

"The top one, Monsieur."

Foucault examined his weapon, making sure that the butt was of just the right thickness and weight, balancing and setting it comfortably in his hand. He seemed satisfied with the color of the metal parts, a deep blue that did not reflect light and would not dazzle him when taking aim.

"The trigger. Monsieur. Is it a hair trigger?"

"No, you will have to squeeze it. We thought it unwise to set it too finely."

Foucault seemed satisfied. Though he was not an excellent marksman, he did come from New Orleans which enjoyed its reputation as the Dueling Capital of the New World. And it was very windy just now. And he trusted Harrah's judgment that this would somehow interfere with Timmons's ability to shoot straight. Besides, Foucault was a gambler, and he had heard that the odds were five to one against being hit when a shot was fired in a duel. And fourteen to one against being killed. Not an unfavorable percentage for a betting man.

Reassured by his calculations even as Timmons removed his pistol from the box, Foucault straightened his back which tended to be round-shouldered and awaited instructions.

"Let's begin," said Harrah. With Foucault at peace with himself, he sensed that this was the best time to proceed with the unpleasant business.

"Will the gentlemen stand back to back?"

Foucault and Timmons positioned themselves.

"There will be three shots in all. I will count six," continued Harrah. "After taking the sixth step forward, you will each turn and fire."

"But where I come from," protested Timmons, "it's usually four steps."

"This is Cape Island, Sir. We do things differently here."

Timmons assented. Just as Harrah began his count a swirl of wind whipped up the sand, blowing it in all directions. Timmons expected Harrah to stop and begin his count again. But he did not and the numbers progressed. At the count of five the duelists were at a distance of some thirty feet from one another.

"Six!" called out Harrah.

The men wheeled about and fired. Timmons's shot missed his man. But Foucault's bullet quite by chance, for he was to much in a hurry to aim properly, blew his antagonist's hat off.

At first, astonishd by the effect of the shot, Timmons was relieved that it did not injure him. But then relief quickly turned to rage. Foucault had inflicted the prime insult. As a gesture of contempt, he had shot off his hat without wounding him.

Hastily reloading, Timmons held his pistol at arm's length and tried to find Foucault in his sight. But the skittish wind and whirling sand made this impossible. He stepped forward, a serious breach in the dueling man's code, and immediately heard shouts of "Foul!" by Harrah. But Timmons refused to step back and, catching a glimpse of Foucault, despite sand in his eyes, fired his second shot. He no sooner heard the report of his own pistol and, hot after that, Foucault's return firing than he loaded for his third shot.

By now Timmons was incensed, unable to control his rage. How did he ever subject himslf to such unfavorable conditions! How did he ever allow this dandified son of a quadroon bitch to survive his first two shots! He rushed forward, stumbling on the treacherous sand, in pursuit of Foucault. But the gambler did not wait to be struck down. Tossing his pistol away, he ran from his antagonist and headed towards the Landing which was in distant view.

As Harrah and Woodbourne gave chase, Timmons doggedly pursued his man. Tripping on a rising sand dune and falling to one knee, Timmons took careful aim and fired at the fleeing figure. This time his shot hit the mark. Foucault toppled as he ran.

Harrah and Woodbourne were appalled. Catching up with Timmons, Woodbourne threw himself on the man, bringing him down, and the two rolled over in the sand. Harrah pushed on until he reached the fallen Foucault. Getting him to sit up, he removed the man's coat and found blood on his shirt.

"How bad is it?" asked the gambler with some discomfort, if not pain.

"A flesh wound, no more," concluded Harrah. "We'll have Dr. Marcy take a look at it."

Foucault sighed relief. A moment later he added. "I think the time has come."

"For what, Armand?"

He twisted his moustache. "For me to leave Trescott House. I've caused you too much trouble already. You'd do well to keep professional gamblers like me out of your establishment."

Harrah did not disagree with the man, but he said nothing.

"At least I'll leave you with a full house," smiled Foucault.

"A full house?" repeated Harrah. "I think not."

"Why not?"

"Mr. Timmons will also be leaving. It's bad enough he breached the rules of dueling. I can't allow him to break the rules of Trescott House, too."

"Which rules, Mr. Harrah?"

"The rules about paying in advance. Thanks to you, the man's three weeks behind in his rent!"

1849

MRS. CULPEPPER

ACQUISITION

Ephraim Eldridge, one of the many Eldridges on Cape Island, brought his Jersey wagon to a halt in front of Trescott House. The wagon was a graceless vehicle with four creaking wheels and square, straight sides that resembled an open coffin.

A scratch farmer, who barely "scratched" a living out of the soil, Eldridge supplemented his meager income by providing rough transportation from the boat landing to Cape Island. Occasionally he hauled furniture for the smaller houses: beds, rockers, tables, dressers, mirrors, wash stands—whatever an unpredictable season called for. Harrah had employed his services on several occasions, though the farmer's performance was never on a par with his price. But he had little contact with the man otherwise.

"What can I do for you, Ephraim?" asked Harrah, his thumbs in his vest pockets.

"Well, I have a favor to ask."

"A favor?"

"Yes, things have not been going too well for me. Mine is not the best farmland. My patch isn't big enough. And wagonning back and forth is a fitful business. And, of course, I've never been one for the sea. So I've been thinkin'."

"Thinking of what?" Harrah asked the question although he knew he would have the answer soon enough.

"I've been thinkin' of movin' on. To Indiana or Illinois. Or even to California."

"To join the Gold Rush?"

"I'm thinkin' about it."

"Then I wish you luck," said Harrah.

"Trouble is, I have no one to stake me." The farmer did not rush his words.

"There are a lot of Eldridges in these parts. Close kin, I imagine. Maybe one of them can stake you."

"A couple of them are movin' on too. You see, Cape May's all right for them that owns hotels or boardin' houses on the Island. But it's hard on the rest of us. Summer helps—what with the boat landing and my wagon. But the season only lasts a few months. And we've got to stretch what money we make to last the whole year."

"I understand," said Harrah. He was sympathetic to the man's plight, but there was precious little he could do.

"You've seen my little house, haven't you? It's the second one past Higbee's Landing Road. About a quarter mile past."

"I don't recall."

"Then let me show it to you. You see, Mr. Harrah, I'd like you to buy it."

Harrah was prepared for a handout, but not this. Buy a house? In the middle of nowhere? The man must be daft.

"I'll let you have it for eight hundred dollars. I need the money desperate. The least you could do is look at it."

Even at eight hundred dollars the house and land did not sound like much of a buy. But as it was still preseason Harrah agreed to take a look.

He climbed on the wagon and drove out to the place with Eldridge. As he had expected, the house was much neglected with farm implements crowding the kitchen and spilling out into the living room. But somehow the cluttered brick fireplace appealed to him and the solid floor. And there was little, if any, structural repair needed.

"Has a hidden room, too," boasted Eldridge.

"Where?" asked Harrah, only passing curious.

"Follow me upstairs."

They took the narrow, low-ceilinged staircase that wound behind the fireplace to the room upstairs.

"It's hardly hidden," remarked Harrah.

"Watch," said a proud Eldridge as he pushed a cupboard to a side. "Thought there was a wall, didn't you? But it's an extra room. I use it for storage."

"Can't use it for much else," said Harrah, noticing that the room had no window to illuminate the trunks and boxes it housed.

"Well, what do you think?"

Even if Harrah wanted to but the house, he was hard put to lay his hands on eight hundred dollars just now. The refurbishing of Trescott House had put a strain on his resources. And he was planning to send Jonathan to Reverend Williamson's Cold Spring Academy in the fall, which was inexpensive enough but entailed other costs.

"Why don't you look for another buyer?"

"I have. No one is interested. And I can't let it go for less than eight hundred dollars."

Harrah tried to find a reason for buying, not so much for Eldridge's sake as for himself. And then it struck him that Jonathan or Willacassa might need a house one day, at least something to start out in. The cottage might not be so isolated in five years or ten.

"When are you leaving?"

"By July—if I had the money."

"All right," said Harrah. "We'll find a lawyer in Cape May

Court House and draw up the deed. But I don't want anyone to know about it just yet.''

''Your creditors?''

''Exactly. If they get wind of it, I'll have no money for your house. I can pay them off in August once the season is under way.''

''You can rely on me, Mr. Harrah.'' Eldridge stuck his hand out to clinch the deal.

MATCHMAKING

The owner of Ocean House paid Harrah a visit, the first time ever. Mr. Leaming had never gotten over his Yankee ancestry. His own people, of course, had been early settlers of Cape May County, the first Leaming coming to Town Bank over a hundred fifty years ago. One Aaron Leaming had purchased all of Seven Mile Beach over a hundred years ago. And there was Leaming Plantation, which Harrah had heard of but never seen, somewhere to the north, a huge land holding which stretched into Five Mile Beach. An industrious, hard-driving, tactiturn man, getting on in years, Israel Leaming still regarded Harrah as a newcomer though he had been living on Cape Island for more than seven years.

"This is an unexpected pleasure," said Harrah.

"You've improved your house, I see?"

"I had young Henry Sawyer working on it."

"Many guests this season?"

"I have no vacancies."

"Good. And a few eligible gentlemen among them?"

"Eligible?"

"Unmarried but prosperous."

"I suppose so. Why?"

Israel Leaming took the seat offered but did not lean back.

"There's a young woman, recently widowed, staying at my place. Actually, her husband died this past August, leaving two children. She's wealthy, not much over thirty, and in need of a good husband—"

"I'm not in the matchmaking business, Mr. Leaming."

"Nor am I, Mr. Harrah. But I've known the family for years. Mrs. Culpepper and her children come to Ocean House every summer."

"So you feel an obligation toward her?"

"Exactly. Is there someone she might be interested in?"

Harrah considered a moment but could come up with no names. "And what if there were. How do you propose to bring them together?"

"The lady is looking to sell part of her estate. We could pretend the gentleman's a prospective buyer and have them meet here. She is very modest and would take offense if she knew."

"I understand," said Harrah.

"Then you will cooperate?"

"I'll send you a guest list if that will help. But I leave the rest to you. I want no part in the matter."

"Very good," said Leaming, rising to go. "One thing. The gentleman should be a Southerner. Mrs. Culpepper is from Virginia."

"That should be no problem. More than half my guests are from Dixie."

Madeleine Culpepper was a woman of thirty-one or thirty-two, as Leaming had described her. But no mention had been made by Leaming of her good looks. Of average stature, she had blond, wavy hair which she wore loosely tied behind her. But her face, to Harrah's mind at least, was of exquisite delicacy and beauty. Cheeks deliciously curved. Eyes clear and sapphire. Lips a perfect, gradual bow. And as modest and ingratiating a smile as Harrah had ever seen.

More than that, she had soft blond shoulders that were bone where there should be bone and flesh where flesh should reign. There was also a hint as she stepped down from her carriage, but only a hint, of a delicious bosom at the edge of her pale yellow gown.

No sooner was Harrah smitten by the apparition of Mrs. Culpepper as she entered Trescott House, his house, than he was jolted by the appearance of her dinner partner who came to greet her. Talley Blatherston was a horse breeder from Virginia who had a great deal of trouble keeping a rein on his self-congratulatory word gallop. Whenever Harrah encountered the man in the parlor or the drawing room, he found an excuse to take quick leave of him.

"He blows like a whale," he confided to his son, Jonathan. "I hate to talk about my guests, but I find the man insufferable. Given his propensity to spout, he should have taken lodging at Old Town Bank. Maybe there's still an ancient harpoon about that could be stuck in his blubber when he starts to blow."

Actually Harrah pointed all this out because Jonathan was somewhat awed by the man. To Jonathan, the word and the deed were one. When Blatherston said he was planning to take a boat around Cape Horn to join the gold rush, Jonathan believed him. When he claimed his horses were descended from those that Saladin had ridden when he frustrated Richard the Lion-hearted during the Crusades, Jonathan never doubted him. And when Blatherston spoke of his prowess with women, unmarried and widowed, the boy was sure that his list of conquests was at least a scroll long.

Blatherston had his better qualities. He was punctual, fastidious about his bills which he paid in gold dollars and double eagles, generous with his tips, and congenial enough to those who paid him homage. But even in casual conversation he left little room for anyone to squeeze in a word. And Harrah was not even about to try.

He did manage every now and then to get a peek in at the dining room to see how Blatherston and Mrs. Culpepper were faring over the white Irish linen table cloth. But he could not determine from her pleasant smile and lady-like bearing how she was taking to the man or what she was thinking. She kept drinking glasses of water which Harrah brought in daily from Cold Spring, Cape Island water being flat and brackish, while her dinner companion poured himself wine which he gulped between his outpouring of words. She did nod once in appreciation after tasting the shrimp cocktail served to her by the young Willacassa. And Harrah began to fear as he returned to his office that unaccountably this apparition, this beautiful woman, like his son, was also impressed with Blatherston. It was enough to make one knock ten-pins down with a cannon ball.

At long last the dinner was over and Mrs. Culpepper was shown to her carriage by Talley Blatherston. From his small office window Harrah could see them exchange pleasantries, Blatherston looking like the whale that swallowed a school of herring. Finally with a sweep of his tailcoat he bade her goodbye.

The next morning Jonathan informed his father that a lady was waiting to see him. Harrah left his office and found a lovely but indignant Mrs. Culpepper standing at the clerk's counter.

"Mr. Harrah. You are Mr. Harrah, aren't you?"

"I am."

"May I have a word with you?"

He could see she was distraught.

"Of course."

"In private," she insisted.

"Will my office do?"

She nodded and followed him to a small room at the rear of the main floor.

Inside, Harrah offered her a chair. "I must apologize for the size of the room," he said, sitting down at his desk. "But space is at a premium at Trescott House, unlike Ocean or one of the bigger hotels."

"I'm not here to chitchat," she said, clasping her small white hands in her lap. "I want to say that I have never met a more insufferable man than Mr. Blatherston. He's loud. He's pompous. And he goes on and on about his horses, his deals, and his masculinity—none of which I care about in the least. But to make matters worse, he did not once question me about the sale of my property. One would think I was here to see if I was a likely filly for him." She looked accusingly at Harrah. "That was not the purpose of the meeting, was it?"

Harrah was too embarrassed to offer a reply.

"Then it's true. I suspected as much, but I didn't think you would have the audacity. Finding a gentleman for me would be bad enough. Matching me with someone of Blatherston's caliber is not only insulting but tasteless. I'll certainly speak to Mr. Leaming about this!" She rose to go, but she was so overwrought by the idea of Talley Blatherston that she began to tear.

Harrah wanted to comfort the woman but thought better of the idea.

"Mrs. Culpepper," he began, "I want to assure you that I had nothing to do with your meeting Mr. Blatherston. He's a guest in my house, so I won't comment on his character or personality. But at no time did I suggest that you and Mr. Blatherston—"

"Would make a couple?"

"Your choice of words, not mine."

"Then whose idea was it?"

"I'm not at liberty to say."

"Oh, you men will stick together!" She sat down again, more to control her sobbing than to suggest any desire to stay. "If you were not responsible, Mr. Harrah. If it was Mr. Leaming's idea—and who else could it have been?—then I suppose I owe you an apology."

"On the contrary, I did provide Mr. Leaming with a guest list. I confess to that. And I did agree to having the meeting take place here. But I had no idea Mr. Blatherston would be selected for the honor. Had I known I would have made quick work of it."

"But why Mr. Blatherston? Why such a bore! He puffs and blows like a giant sperm whale! Doesn't Mr. Leaming have any regard for me?"

"He has the utmost regard. I imagine he selected Mr. Blatherston purely on account of his residence. He does come from Virginia, you know. Just as you do."

"The pity is Virginia's then." Mrs. Culpepper squeezed the skin around her eyes and choked a breath through her tears.

"It's bad enough being a widow this past year. But I might survive it if people weren't so anxious to have me marry again. Don't they understand? I don't want to marry again. I want only to be left alone."

Harrah reached into his drawer and pulled out a clean handkerchief. "Try this," he said. "Yours is about done."

In spite of herself, Madeleine Culpepper let a tiny laugh break through.

"I suppose it is," she said, putting her own handkerchief down. "I'm sorry to be making such a fuss. I'll be all right in a moment."

"There's no hurry," said Harrah, finding even greater beauty in a distressed Mrs. Culpepper than he saw in her before.

"Will you fetch a carriage for me?" she asked, more herself once more.

"Certainly," said Harrah reluctant to be leaving her but getting up just the same. "I'll be back in a little while."

When he returned, Mrs. Culpepper was again her composed self. Except for a little residual redness about the eyelids, one could not tell she had been crying. She said nothing to Harrah as he led her out the back passageway and through the small garden that Rachel tended in spare moments. At the carriage he halted and offered Mrs. Culpepper a hand up. She took his hand and settled in her seat. But she said nothing as the driver was directed by Harrah to Ocean House.

The next morning a Negro messenger boy delivered Harrah a note.

"I'm so sorry about yesterday. My behavior toward you was inexcusable. Mr. Leaming confessed all. He even acknowledged your reluctance in the matter. The more's your credit that you did not know who the lady was. I do hope you won't think ill of me.
                                                Madeleine Culpepper"

Harrah's first impulse was to let the matter rest. Everything was patched up and his reputation was restored. But the letter gave him no rest. It called for a response. Even if it did not call for one, it provided the opportunity for a reply.

He held off for several hours, then sent the following note:

"Even though I have been exonerated, I am not without guilt. Let me make amends. Let's have dinner—on neutral ground— neither at the Ocean House nor at my place. I promise not to question you about the sale of your property. And I will not urge you into remarriage. Will tomorrow be too soon?
                                                Nathaniel Harrah"

Her reply came within the hour.

"Much too soon. Make it Friday instead."

MAKING AMENDS

Early Friday afternoon Harrah picked Mrs. Culpepper up at Israel Leaming's Ocean House on the east side of Perry Street. Ocean House, opened in 1840, stood near Congress Hall. Large, commodious, undistinguished, it accommodated two hundred fifty people. But to Harrah the other two hundred forty-nine were superfluous.

Madeleine Culpepper was ready and waiting for him when he drew up his carriage. She wore a yellow dress with a modified hoop skirt so that she would not have too much trouble sitting down.

"Where is 'neutral ground'?" she asked.

"Neutral ground? Oh, yes. I'll show you."

It proved to be what was once a private residence on Washington Street, which now served dinner. With space for only four tables tucked away in different niches of what passed for the dining room, the little restaurant was as romantic a spot a couple could desire. The wooden tables had no cloth but were covered with a kind of fish net that created both an airy and nautical feeling. The only light, however, was provided by ship's candles that quickly used up the oxygen in the room, leaving the occupants as drowsy as wine drinkers.

"Wherever did you discover this place?"

"There's very little about Cape Island its inhabitants don't know."

"Excellent!" she proclaimed as she took a spoonful of the bowl of hot clam chowder that had been set down before her by the owner of the house and cook, a Mrs. Corgie.

Warmed by the soup, Madeleine Culpepper looked about the room they were in and the three empty tables in its other corners and asked, "Why did you ask me to dinner?"

"To make up for Talley Blatherston."

"Oh, that man! How do you stand him—even as a guest?"

"I avoid him as much as possible."

"It's strange," she said. "I've been to Cape Island for several summers. And I never knew Trescott House existed."

"You weren't alone in that regard. Until last year it was in imminent danger of going out of existence."

"It's really quite luxurious compared to Ocean House."

"We cater to a different clement now. The wealthy and those with class prejudices, sometimes called pride."

"Like Talley Blatherston!"

Harrah chuckled. "Well, something like that. But not as terribly afflicted.''

"If you find your guests so objectionable, why do you rent to them?''

"They're not all objectionable. It's just that the few that are seem the rule rather than the exception. Besides, it's a livelihood.''

"And you have children to raise?''

"Yes, two. Willacassa waited table on you the other day.''

"A lovely girl.''

"And Jonathan showed you to my office when you came looking for blood.''

"I'm sorry for that.''

"I've no complaint. We wouldn't be here today if you hadn't come.''

Madeleine Culpepper accepted Harrah's explanation without comment. And he could see that she was as modest and gentle a person as she was physically lovely.

"I have two children, too,'' she informed him, stopping only to marvel at the platter of seafood that Mrs. Corgie placed before each of them. Such an array of oysters, clams, scallops, fried shrimp and filet of sole she had not seen before, even on Cape Island. And the aroma of the butter sauce and sautéd slivers of green onions tickled her palate.

"What a gorgeous ensemble!'' She sampled the scallops and the sauce and pronounced them perfect.

"But I do worry,'' she continued, "what I will do when the children are grown. I know that they will have families of their own. And their own interests. And I wonder how I will manage without them. Indeed, I sometimes wonder what life itself is about. It seems so chancy and accidental. I suppose I feel this way because I lost my husband. But when I look at other people, I find that they too are tossed about on a sea of circumstance.''

Harrah was not quite prepared for this. The women he had come in contact with on Cape Island were invariably on holiday. They thought only of the pleasures of the resort and of drinking in the sea breezes. Georgina, of course, had a more practical turn, given her husband's illness and impending death and the long-range concerns arising from it. But no one so thoughtful and so vulnerable as Madeleine Culpepper had crossed his path.

"Life is chaos,'' he observed. "That's the terrible truth of it. The whole point of life is getting order out of chaos. Creating strategies and forms and patterns. And hopefully beauty and new directions.''

"But why? It seems so futile.''

"To give it meaning. Without meaning, life would be intolerable.''

"Is that why you built Trescott House?"

"I didn't build it. I bought it. But I did alter it somewhat."

"What do you do when summer is over? I imagine it's pretty desolate then.''

"Desolate? Not really. There's always the hurricane season. In late September or early October, I stand at my window and watch the waves smash against the shore or swallow up the beach. Sometimes a ship is caught in a storm. And if it's wrecked, we run out on the sand to pull in survivors.''

"High drama.''

"Exactly.''

"But there must be quiet times," she smiled. "Times when the sea is calm and no ships are in distress.''

"Yes, but there's always work to be done. Repairs on the house. Painting. Refinishing and refurbishing furniture. Storing in provisions for winter. Bringing in wood for the fire and oil for the lamps. Caring for the horses.''

"How cold does it get in winter?''

"Quite cold. But more than that, it's raw. The wind carries a lot of moisture. And it has a biting edge to it when angry.''

"But surely there must be more to your life than work and chores.''

"Yes. I have a few friends. The Sawyers—he's a carpenter who does work around the house—other hotel keepers, Reverend Williamson who hasn't yet succeeded in getting me to church. And, of course, there are the children. Jonathan is fourteen. Willacassa will be twelve in December. They've reached the age where I can talk to them and find them as keen and exciting as they are entertaining.''

"You're partial to Willacassa, aren't you?''

"I suppose I am. But Jonathan doesn't resent it. He's my son and I expect a little more from him than I do Willacassa. This sometimes puts a strain on the boy. But he's up to it for the most part.''

"So being a parent and a businessman you have time for little else.''

"Oh, for a brief period in my life I made a living as an artist. Not much of a living, but I tried. I still do sketches and paintings now and then.''

"Not of nudes I hope.''

Harrah smiled at her daring.

"Not lately.''

"You must show me your work some day," said Madeleine

Culpepper with unfeigned interest.

"I'm afraid that won't be possible. I want you to have a high opinion of me. And I'm not generally that good. But I will show you Cape Island. It's a somewhat more impressive canvas than anything I've done."

"I think I would like that," said Mrs. Culpepper.

"Must I wait till next Friday?"

"No, would tomorrow be too soon?"

"Tomorrow would be ideal." said Harrah, raising his wine glass almost as a toast.

A MUTUALITY OF UNDERSTANDING

Almost from the first Harrah knew that Madeleine Culpepper would soon be the dominant woman in his life. She did not fill the void Cassandra's death had left. No one could take Cassandra's place or dim his memories of her. But there was no bringing Cassandra back! If he was ever to put the old life behind him and start over again, this was the person to do it with.

Had someone told Harrah that Madeleine Culpepper and he "understood one another," he would have laughed at the idea. How could two people from two different worlds understand one another? Madeleine was gentry from Virginia. And he was a transplanted New Englander who had come to Cape Island by way of Philadelphia and the Pines. Yet, there was no denying it. There existed a mutuality of understanding between them that made explanation unnecessary and quarrel unlikely. And he could see himself spending the rest of his days with her, cherishing every moment of their time together.

There was, of course, one difficulty. An unbridgeable gulf stood between them that even their mutuality of understanding could not span. It was not so much a geographic gulf—Virginia was not that far away from South Jersey—as a psychological gulf. A gulf created by the conflicting times they lived in and by what Harrah regarded as the unforgivable blemish on the face of midcentury America. The gulf, of course, was slavery.

It was not that Madeleine Culpepper advocated slavery. He did not know what thoughts she had on the subject. The topic had never introduced itself. The things that Madeleine Culpepper talked about were her children, her life in Virginia, and Cape Island, the Cape Island that Harrah had made a point of showing her. When she had come to the island before, on the arm of her husband, all she knew of the place was what any vacationer knew: the names of a few hotels and boarding houses, the beach, the bathhouses, the ocean, the churches, and the occassional store that catered to the season and the season's guests.

She knew nothing of the pirates who had frequented these shores in earlier times. And she was surprised to learn that the inhabitants still believed Captain Kidd had buried treasure there, possibly near the very same cedar tree that Harrah had pointed out. She had only a fleeting awareness that the original Town Bank, which was now buried in Delaware Bay, had been a whaling village and the first settlement on Cape May. She did not know that Cape May Court House was not a courthouse on Cape Island but the county seat some dozen miles to the North. Nor did she know that Daniel Hand who was now working on the new courthouse was the same man who had renovated Trescott House. And she was amused to learn that the house built by Thomas Hughes in 1816 and originally called the Large House, was renamed Congress Hall when Hughes became the first man elected to Congress from maritime Cape May County, amused because when Hughes rose to speak for the first time, he was told, "Sit down, Clam!"

All this Harrah told her or showed her, just as he pointed out Higbee's Beach and the red brick church at Cold Spring with its old tombstones, their names rubbed away by time. He even pulled out a whaleboat one day at the east end of the island and rowed her all the way to Five Mile Beach, a wild, wooded area thick with cedar, oak, and storm-twisted holly. Near a grove dripping with Spanish moss, they picnicked on the sand and watched an army of gulls marching about the shore, like new recruits on a disorganized drill field.

"Cape Island has a lot of growing to do yet," said Harrah as they ate their cold chicken and huddled together on the windswept beach. "But I wouldn't be surprised if Five Mile Beach and stretches of land to the north also housed summer guests one day. All it has now is an occassional fisherman. But there's a lot of sand in these parts. And where there's sand, sky, and ocean, there will be summer cottages and summer visitors."

As he talked, Madeleine Culpepper, her hair windblown, turned to Harrah with as loving a smile as he had ever seen. And for one brief moment this expression of sentiment fired in Harrah feelings of affection, even passion, he had not known for years. Cape Island was the Autograph of God. But so was the beauty of Madeleine Culpepper! Should he seize the moment? he wondered. Why not! He leaned over and kissed her.

Madeleine Culpepper did not draw back. If anything, she was captivated by the man and savored the kiss. And as his lips slowly pulled away she knew that though Harrah in no way reminded her of her late husband she wanted to share her life with him. For Harrah's quiet strength reassured her. And she trusted him, not as a woman in love for many years trusts a man but as a woman who falls in love because of such a feeling of trust.

And so she placed her children almost totally in the care of the governess. And she felt little guilt over this because she wanted as much time to see Harrah as she could find. After all, there were but a few weeks left to the season. And the children had the sea, sun, and sand to occupy them. As long as their mother was there at breakfast and at dinner, and in their room to put them to bed at night, they had no complaint.

For his part, Harrah grossly neglected his duties at Trescott House, leaving the establishment to function pretty much on its own. He had no designated assistant. And what Rachel could not do, Jonathan would not do. But Willacassa somehow got it done—unless she was able to put the matter off until Harrah made one of his rare appearances during the day.

What Harrah and Madeleine Culpepper did when they were together was neither exotic nor extraordinary. They walked on the strand in the cool shadow of late afternoon. They took a turn in his carriage. They stopped to look at some of the shops on Washington Street or drank iced tea at an open window. But they were in each other's company. And that, magically, made all the difference. Nothing exotic was needed, not even the glorious sunsets. For Cape Island itself was exotic, a place far from the maddening concerns of everyday life.

But as the season drew to a close, Harrah felt the pressure of time. In a matter of days, Madeleine Culpepper would board one of the steamboats at the Landing, her children at her side, and their moment together would be over. He tried to think of some way to pursuade her to stay on.

"The weather is at its best in September. Clear skies and cool. It would be a shame for you to leave before then."

"But everyone's leaving."

"Creatures of habit."

"Creatures of responsibility, you mean."

"I want you to stay," he said softly.

"And I want to stay," she replied. "But I can't."

"Not even for me?"

"It has nothing to do with you. You must know that."

"If you stay, I'll let your children have the run of the place. Trescott House will be theirs to do with as they please."

"Are you proposing marriage?"

"No," he lied. "You said you wanted only to be left alone. And I respect that. But does it mean you want to be lonely?"

"It does get lonely at times—even with the children."

"I lost someone too. It's been eight years now," said Harrah. "And I've known my share of loneliness. I don't mind it most of the time. There's work to do. And the children help. There are times when I want to hear a woman's voice. Not just any woman's voice. But the voice of Madeleine Culpepper."

"Then I've nothing to fear?"

"Nothing."

"I was going to say, 'Good.' Now I'm not so sure."

"Just say you will not be leaving with the others," said Harrah suddenly. "I'd love for you to stay on a while."

"I can't, Nathaniel. The children must get back. There is their schooling to think of. And the Culpeppers will want to see them. My own family—what's left of it—is small and living in Mississippi now. The Culpeppers are all the family my children have."

Harrah lowered his eyes, astonished at the pain this caused him.

"Don't look so sad," she whispered. "I'll be back next year. I promise."

She squeezed his hand to keep her emotions in check. "Besides, if I stayed, you'd want fulfillment. And I'm not prepared for that just now."

Harrah inhaled the strong ocean air.

"But you will be back?"

"Yes, my dear. We'll see each other again. If you love me, as I do you—Yes, Nathaniel, I love you and I admit it freely—it won't be so long to wait."

1850

FUGITIVE CLOUDS

MIDCENTURY

As predicted by Harrah, midcentury on Cape Island began with a building boom. Impressed by the success of such old stand-bys as The Mansion House, The Atlantic, Congress Hall, Centre House, Ocean House, Washington House, American House, Cape Island House, and the newer Columbia, New Jersey, and Franklin Houses, speculators began the building of new hotels in earnest. The Tremont House, on the corner of Washington and Franklin Streets, was opened by Humphrey Hughes, Jr. Aaron Garretson's Nathaniel Hall was not only built but painted white. "A sign of things to come," suggested Dr. Samuel S. Marcy who built his own hotel, White Hall, on Lafayette Street above Jefferson. Philadelphia House, which was opened by H.D. Stuard next door to Cape Island House, gave Lafayette Street and offshore Cape Island a new prominence. There was even a new boarding house constructed at the steamboat landing on the bay where, Harrah heard, refreshments were served at all hours.

But it was the United States Hotel which was the handsomest of them all. Presenting two fronts on the southeast corner of Washington and Decatur Streets, both over a hundred feet in length, the hotel surrounded four of its five stories with ornate but graceful pillars and verandas. And at the top of the hotel, an airy dome-shaped observatory, with flag flying, pierced the billowing sky.

Not to be outdone, James Mecray enlarged the Delaware House which had been built ten years earlier. And Lilburn Harwood of Columbia House, across the street from Harrah (where Georgina Brookens had stayed), added a large south wing, five stories high, that doubled the capacity of that hotel.

Harrah was not too happy that Columbia House, now the largest hotel on the Island, was next door to him or that the magnificent new dining room, some one hundred twenty feet long and thirty-five feet wide with a ceiling fourteen feet high, was lauded as having the best cuisine at the resort. Harrah could compete with quality before. Now he was not so sure. Rachel was an ingenious cook with fish and meat, but she had her limitations with garnish and condiments and exotic salads. If Columbia House was not only the largest establishment on the Island but could outdo in luxury and the culinary arts what little Trescott House could do, then the only thing that stood between the two was gambling.

But Harrah need not have worried. Despite the building boom, the resort was so overcrowded at the start of the season that every available cottage, every room, every bed was taken over by the influx of visitors. Dining halls, parlors, vestibules, sheds, tents, even outhouses and open boats were taken over for emergency sleeping quarters.

"Give me a rug and I'll sleep on the floor," pleaded one visitor.

"Try the smaller cottages or boarding houses two miles out of town."

"I've tried them."

"Then a fisherman's shack, whatever you can find."

"Someone said that you have a large room that's still unoccupied. I'll give you twice the going rate."

"I'm sorry," said Harrah. "It's reserved." He refused to rent out the room he was holding for Mrs. Culpepper.

But the season had already opened and Mrs. Culpepper did not appear. A lively Fourth of July, celebrated with fireworks and booming cannons and parties and the ringing of church bells, passed into history. And still no Madeleine. She had written she was coming. But when?

Then news reached Cape Island that General Zachary Taylor, President of the United States, had died of cholera morbus on July 9. Cholera morbus was a common illness in the hot months, common enough for ordinary men with its vomiting and purging. But why a president? And why "Old Rough and Ready," hero of the Mexican War?

Cape Island went into mourning. Flags were flown at half mast and black crape was hung from the verandas of the big hotels. The same day, steamboat passengers reported a disastrous fire had broken out in Philadelphia. The toll was devastating, three hundred fifty-four houses and scores of lives. Large numbers of Philadelphians had always flocked to the Island, and the town was properly solicitous of their feelings. Church services were held, dances were cancelled, and a moment of silence was called in dining rooms prior to serving the meals. Gloom greeted the new arrivals.

But Harrah's gloom was heavier than most. He had waited all year for Madeleine Culpepper. He himself had been down to the steamboat landing the whole first week of July. During the next few days Harrah sent a carriage and a wagon on the chance she would surprise him. It began to look as if she would not come at all.

"Why are you so cross, Daddy? Aren't you happy that we're all fill up?"

"Of course, Willicassa."

"Well, you don't look happy."

Harrah realized that he need not wear a long face any more than a heart on his sleeve. He apologized for being unapproachable and gave Willacassa a big kiss.

"Do you have one for me, too?" Madeleine Culpepper was smiling at the entryway, having just arrived.

Harrah ignored his patrons and took the lady in his arms. "Where have you been?" he asked. "What delayed you? And where are the children?"

"Give me a moment, Nathaniel. I've just gotten here."

"I'm sorry. It's just that—"

"Why don't we step into your office for a moment. How are you, Willacassa?"

"Just fine, Mrs. Culpepper. It's so good to see you again." Willacassa was too grown up to kiss her, but her pleasure at seeing the lady was unrestrained.

In his office Harrah ended the long famine of wanting her. He kissed Madeleine and stroked her hair and pressed her hand in his. Then with the understanding that they had passed the first stage of their relationship, he drew her towards him and gloried in the touch of her breast.

"You did miss me then?" she asked, warming towards him.

"I was half out of mind when you didn't arrive."

"And I missed you, my love. This has been the longest year of my life—this waiting to see you. I can't believe the wait is over. I am so happy the summer is ahead of us."

"Minus ten days," teased Harrah.

"I had to attend a funeral in Mississippi. An aunt who was my closest kin wanted to see me before she died. She was the last of the Harboroughs. She died in June, poor thing. But I just couldn't leave. You understand, don't you?"

Harrah nodded.

"I wrote a letter from my cousin's place. But I suspect I arrived before the letter did."

"Mail is a problem here."

"One more thing," she said, slipping her breast back into her gown. I'm not staying at Trescott House. I've taken my old room at Ocean House."

"Ocean House?"

"Yes, Mr. Leaming met me at the steamboat landing. He inquired why I wasn't staying at his place this year. I told him I didn't think he would have a room at this late date. But he did have one and switched people around so it was the same as before. I thought it best not to tell him about Trescott House. I don't want to raise any eyebrows. Besides, the children could make things awkward for us."

Harrah was disappointed, but not so disappointed he could

not see the wisdom of her action.

"Then your children were taken to Ocean House?"

"And my luggage, too. You're not angry with me, are you?"

"I'm not happy," admitted Harrah. He squeezed her hand. "But I'll find solace in knowing you're on Cape Island. There's a good deal of solace in that."

OUT OF THE RAIN

He took her to what used to be Ephraim Eldridge's house. Ostensibly it was to get out of the rain. But Madeleine knew what he had in mind once he managed to push the door open.

"No need for a key," remarked Harrah. "Not with this door."

"Whose house is it?"

"Mine."

"Yours. I thought you had better taste than that."

"Better taste, yes. But not the money to go with it." He dusted off the couch. "Actually I bought it as a favor. And maybe one day I'll fix it up and give it to Jonathan or Willacassa as a wedding present."

"At least it's dry," observed Madeleine, glad to be out of the downpour.

"It'll be drier still once we get a fire going. Better take your clothes off."

"Mr. Harrah!"

"Upstairs, of course. You may find some old workclothes in the bureau."

"Workclothes! What will you think of me when I come down in workclothes!"

"You'll be just as fetching, I'm sure."

"I'd rather stand by the fire."

"Do as I tell you. I won't have you spending the rest of the summer with sniffles. Besides, it'll be a while before I get the fire going."

She went under protest. By the time she returned downstairs Harrah had gotten a good burn in the fireplace.

"Not so fetching after all," she commented when he turned to greet her. "Am I?"

What Harrah saw took him back for a moment. Madeleine Culpepper was lost in his shirt and trousers, even with their legs rolled up. And her hair had fallen down and half covered her eyes. She could have passed for a chimney sweep—except for the pretty face, pretty even under the most unflattering of circumstances.

"Give me some time," he teased. "And I'll get used to it."

She approached the fire and extended her hands, at least as much of them as escaped his sleeves.

"I can't get over at how heavy the rain was."

"You've got to see it during the hurricane season. I'm constantly warned that Trescott House is too close to shore. That one day the ocean will come through my window."

"Doesn't that frighten you?"

"I guess I don't really believe it." He took her arm. "One thing frightens me though."

"And what's that?"

"That you will go away again for a whole year."

She turned away from the fire and looked up at him.

"Let's not think about that. Let's enjoy our time together." She paused a moment, then broke into a penetrating smile. "I'm ready for you, Mr. Harrah—if you're ready."

"In those clothes?" he asked sheepishly.

"You can take them off."

"After you've gone to all that trouble to put them on?"

"Then I'll take them off myself," she said, sitting down on the couch. And she did, but not without his help.

To Harrah their coming together after so long a wait was more than satisfaction, it was idealization. He had so wanted Madeleine Culpepper that he could not see her as she was, a lovely woman with berry-edged fullness in the chest and a patch of hair at the loins. No, he saw her as a celestial being come to Cape Island to tempt him and to bathe him in rapture. And he could no more deal with her matter-of factly than he could keep from touching her— even long after his passion was spent and he had withdrawn from her.

It did not matter that she had known another man or had borne children. (He did regret that they were not his children.) He saw her virgin and innocent, as if for the first time. And in a way she was innocent, purged by her love of other memory and other times, and born again fresh and pure for Harrah alone.

"I love you, Madeleine Culpepper."

"And I love you, Nathaniel Harrah."

"If you wait a little while, I'll come close to you again."

"You're close now."

"Even closer."

"I'd like that," she whispered.

SCIENTIFIC PROOF

They stopped at the New Atlantic and were lucky to find a table in the crowded dining room.

"Yes, I have a table for you and your lady, suh," said the colored headwaiter, a man called Henson. With as elegant and dignified a manner as Madeleine had ever seen in a man, he proceeded to lead them to their table and adjust their chairs to make sitting down easier.

A moment later when Harrah had given his order and was asking about Madeleine's children, the two became aware of a fuss being made at the other end of the room.

"What do you mean, there's no room? I want a table, and I want it now!"

Everyone turned to see the headwaiter being verbally accosted by a man who identified himself as "A Southerner who has no patience with house niggers who put on airs!"

The headwaiter paled two shades of color, but did not lose his dignity.

"If you will kindly wait, a table may become available, Suh."

"Don't 'Suh' me. Just find a place for me and my companion to sit down!"

It was apparent the man was slightly inebriated and needed to sit down. His red neck was all the more crimson for his rage. When his campanion tugged at his arm in a plea for restraint, he pulled away. "Let go, Dr. Gibbs. I won't be put down by a nigger!" In the act of pulling away, he jostled the headwaiter who had turned to go. Henson stopped short, then jostled him back.

"Damn!" yelled the Southerner, whipping out a pistol from his coat pocket.

At the sight of the pistol, Henson's waiters gathered round to protect him. Three of the waiters flashed knives. In response to this, the Southerner's companion also drew his pistol. A brief scuffle ensued, followed by shouts and death threats. At this point a group of diners intervened and the incident was quickly defused.

Harrah was terribly distressed by what he saw, particularly as the "Southerner" turned out to be one of his house guests, a man who had checked out only that morning. Fortunately his companion, Dr. Gibbs, was a generally calm and reasonable individual who led his friend away.

Harrah made a noble effort to hide his distress from Madeleine Culpepper. But she could see something was wrong. She did not press him about dining together the next day when in an agitated frame of mind he saw her to the Ocean House.

"Good night, dear," she said, kissing him on the cheek. "Don't think too badly of us."

Harrah kissed her in return but allowed nothing more than a few mutters in response.

The morning after this, Harrah had breakfast in his room. When he was sufficiently calm, he ventured downstairs into the parlor. A half-dozen of his male patrons were sitting quietly in armchairs, a chastening silence filling the room. Only a few smoking segars disturbed the morning air, most of the havanas resting precariously on silver-lipped ashtrays.

It was only after a while that Harrah realized his guests had fixed their attention on some amusement at hand. There was a strange silence as what seemed to be pictures, pictures set in pistol-sized cases, were passed around. Harrah could not quite make out what the pictures represented. But to see the men's faces expressionless as they squinted at the cases through the rising smoke—except for an occasional curious sneer—surprised Harrah. It was as if there were a tacit agreement among the men that nothing would be said about what was seen. Yet there seemed to be a general understanding that the subjects at hand were of considerable interest, even significance.

Harrah went about his business, checking the cleanliness of the room and seeing to it that ashtrays and spittoons and newspapers were in their proper places. With daylight filtering through the sheer curtains at the windows, he saw no problem with visibility. He was about to leave when Cameron Gibbs, the doctor from South Carolina, caught his sleeve.

"Have you seen these, Mr. Harrah?"

"I haven't, Dr. Gibbs."

"You should. I think they will be of interest to you. You heard of Louis Agassiz, haven't you?"

"The Swiss Zoologist?"

"Yes. Well, Agassiz himself commissioned these. The gentlemen that have seen them have been most impressed."

Harrah took the first case passed to him and opened it. What he saw was one of the newer daguerrotypes. Not unusual in itself. Except that this time it portrayed not a socially prominent personage who dutifully sat for his portrait but a black plantation slave. A little handwritten note pinned inside to the left half of the case identified the slave as "Jack," a "Driver," from the "Plantation of B.F. Taylor, Esq."

"The photographer does a good job, doesn't he?"

"The photographer?"

"Yes, J.T. Zealy of Columbia, South Carolina."

"But why—"

"But why a daguerrotype of a slave? Have a look at the other cases."

The first daguerrotype had surprised Harrah with its subject, particularly as it was framed with the same ornate mat that was standard for "eminent" personages.

Harrah was not surprised by the second and third daguerrotypes, he was shocked! The second subject, an elderly field hand, also of B.F. Taylor's plantation, "Edgehill," was of such a bitter countenance and emaciated body, photographed naked from the waist up, that Harrah was almost apologetic for looking at it. The slave before him, if Harrah needed any reminder, was a man. But a grizzled old man, so downtrodden and at the same time so intense and defiant, yet bitterly resigned to his fate, that Harrah had pity not only for him but the whole human race, plantation owner, slaver, abolitionist, all men and women born in this century. What humiliation there must have been in being photographed became the humiliation of the beholder of the photograph.

Haunted as he was by this subject, he was ashamed and embarrassed by the next. The elderly fieldhand's daughter, Renty, her hair cut short like a boy's, sat on a chair, her body naked from the waist up. The girl's eyes appeared glazed both from reflected light and movement during the exposure. The girl's face, not quite sullen, was slightly startled. Her fieldhand's shoulders and arms framed her breasts which were not unlike a white woman's except that they were coarse at the nipples, with a wart or growth beneath one of them.

Harrah avoided the doctor's smile and quickly turned to the next daguerrotype.

This time he saw the naked back-view of a standing male Negro, a man identified only as "Jem," a "Gullah" slave owned by F. W. Green of Columbia, South Carolina. As depressing and humilating as the other photographs, this one somehow seemed so pointless that Harrah quickly closed its case.

"What's the purpose of all this?" he asked, unable to shake the feeling of depression he was thrown into by the photographs.

"Scientific proof," asserted the doctor.

"Scientific?"

"Yes. I told you, Louis Agassiz commissioned the studies."

"For what reason?"

"To establish their difference from other races. There can be no doubt now that Africans are a distinct species, and an inferior one at that."

"And Agassiz believes that?"

The doctor was bewildered by Harrah's sudden antagonism.

"He believes in the doctrine of separate creation. He believes the black and white races did not originate from a common center, a common pair."

"But that doesn't make one race superior to another."

"Look at those daguerrotypes, Harrah. Are those men equal to ours? Is that girl equal to our Southern ladies?"

For the first time in his public life Harrah ventured a totally honest opinion. But he did not wait for a reply.

"On evidence I can't speak for the men. But apparently the Negro girls are. How else did we get so many thousands of mulattoes?"

VISION OF A MOTHER

In Madeleine Culpepper, Willacassa had a vision of her mother. Not Cassandra whose charcoal portrait, which Harrah had done from memory, hung on an inside wall of his bedroom, but the mother she would have as a friend, someone to admire and even imitate. Like Willacassa, Madeleine's hair was blonde, but it seemed of a finer texture and was worn shorter than Willacassa's, prompting the girl to get Rachel to cut hers.

Willacassa was already Madeleine's height, but still she looked up to her, admiring the gracefulness of her figure and her seemingly effortless carriage. Willacassa was cut on a large scale, no less feminine and no less attractive. But what others considered striking in the girl—her flamboyant color, for example—Willacassa pooh-poohed, wishing for the same subtlety and reserve that characterized Mrs. Culpepper.

Most of all, Willacassa admired Mrs. Culpepper's placidity and even-temperedness. Never did the gentle lady give way to a display of temper, which was a trait Willacassa exercised frequently at the expense of her brother or any of the boys at school who excited her displeasure.

But it was the way Madeleine Culpepper talked to her, with only a slight Southern accent and a soft, crystal-like voice that drew Willacassa so strongly to the woman.

"My, you're wearing a pretty dress today. It brings out the blue of your eyes. Such pretty eyes you have, Willacassa."

"Thank you, Mrs. Culpepper."

"I would so love for you to have tea with me. I'm sure we can find something refreshing, something cool and sweet to drink, but we'll call it tea just to please the waiters. They are so proper, you know."

"Where'd you get that beautiful locket, Mrs. Culpepper?"

"An old aunt gave it to me. Would you like to wear it sometime?"

"May I?"

'Try it on now. Let's see how it looks. My, but you are getting to be quite the young lady. How well it goes with the color of your skin and—may I say the word?—bosom."

Willacassa colored at the reference but was filled with pride at a grown woman's recognition.

"Would you like to wear it tonight? I hear you're going to the Kursaal?"

"No, thank you, I'm going with my brother. The locket would

be a total waste on him. I'll ask you for it another time. Would
that be all right?''

"Anytime."

"Are you going to see Daddy again?'' This was the greeting
Willacassa gave Madeleine Culpepper one afternoon as she entered
Trescott House. But the girl was so direct and so artless that Mrs.
Culpepper was more amused by the question than made defensive.

"Why, yes, Willacassa. We're going to have dinner together.''

"I'm glad. He's such a slavedriver when you're not here. 'Do
this. Do that. Be sure the silverware is clean.' He has no time to
be a father. Not like the off-season anyway.''

"You like him better off-season?''

"Oh, there's no comparison. Except when you're around.
When he knows you're coming, he's all ice cream and sugar can-
dy. 'Here, Willacassa,' he'll say, 'go buy yourself a new trinket.'
Or he'll come up with tickets to the Kursaal. Straight from Simon
Brolasky himself. We saw a minstrel show from Philadelphia the
other night.''

"You and your brother?''

"Yes. Though Jonathan left for the billiard parlor halfway
through the evening.''

"How thoughtless of him.''

"I didn't mind. A girlfriend of mine took his seat and we had
a fine time. But where are you going today?''

"I don't know. I leave these things up to your father.''

"Daddy will think of something nice. He likes you. And he'll
go out of his way to please you. And, I might say, I like you, too,
Mrs. Culpepper.''

"Thank you, Willacassa.''

She wondered how pleased the girl would be if she knew
where her father was taking her once they had dined.

OBSTACLES

There was no misunderstanding about what a visit to Harrah's cottage meant. But it was a tacit understanding, never mentioned by either one of them. Harrah simply drove his carriage to the cottage after they had dined together and Madeleine Culpepper talked about the pleasant afternoon they had shared. Once inside, Harrah did his best to make her comfortable. He propped up the pillow of the couch and brought out a bottle of wine which he set down with glasses on a small circular table. After they had talked a while and he refilled the glasses, Harrah leaned over to kiss her. It was only then that Madeleine broached the subject.

"You want to make love, don't you?"

"What makes you say that?"

"I know, that's all."

"It can wait," said Harrah, showing commendable patience.

"No, it can't," she said. "I can tell."

"But if you're not ready—"

She started to disrobe, and even before she was free of her gown he took hold of her breast which was cool to touch. But Madeleine was warm and yielding elsewhere. With his help she managed to wriggle out of her clothes, though she would not totally remove her undergarment.

"I ask only one thing," Madeleine whispered as she submitted to his kisses.

"What's that, my dear?"

"Interrupt at the end."

"Interrupt? But why?"

"I don't want a third child just yet, Mr. Harrah."

When she lay in his arms after their lovemaking, she ran her fingers along his chest.

"You see," she said. "I don't want to be alone any longer. I like being close to you. And I'm not frightened of marriage any longer, either."

"I'm glad."

"Are you really?"

She looked up at him, almost accusingly.

"I'd have thought you would have proposed by now. Going to bed with someone is a kind of commitment, isn't it?"

"I suppose it is."

"Unless you're casual about whom you make love to."

"I'm not casual about you, Madeleine. I think you know that."

"Then why are you silent on that score?"

"I may be silent. But I've been thinking about it all the time."

"And?"

"And I'm still thinking."

"Why haven't you proposed? No, really, let's be honest for once. Is it the children?"

Harrah shook his head.

"Don't you love me?"

"You know that I do."

"Then why?"

"Because you have a plantation in Virginia and I a guest house in Cape Island."

"That's silly."

"Is it? Will you live on Cape Island?"

"I would. But I want my children to grow up in Virginia."

"You won't give up the plantation?"

"I would give it up gladly. Frankly it's a burden for me. It's too large for me to handle. I know nothing of crops or management. There are just too many people to worry about. But I don't think I have the right to give it up. I don't think I have the right to deprive my children of their heritage. The Culpeppers have lived in Virginia for almost two hundred years. Your children don't have so much to lose."

Harrah's face darkened.

"I didn't mean it that way. I would never use wealth as a measure or yardstick. You know that, Nathaniel."

"What did you mean?"

"I meant that they weren't born on Cape Island. They have already moved once. Their roots aren't that deep. Another move would not be so trying for them, as trying as it would be for my children. Besides, *you* could manage my estate. I've been looking for someone I could trust. With you in Virginia we could all lead a normal life."

"Is there a normal life in Virginia?" asked Harrah.

"What?"

"With slaves everywhere, can anyone lead a normal life?"

Madeleine turned away. He had evidently touched a raw spot.

"I can't make a brief for slavery," she said at last. "I hate it as much as anyone does. Not only for the slaves themselves. For the Southern women as well. They've got to put up with black brothels and virtually live with a kind of prostitution. It's sickening and degrading. It makes a mockery of Southern society. But it's a fact of life—"

"At least life in the South."

"Yes, and there's no wishing it away."

"There are some Southerners opposed to it."

"I know that. And I'm not arguing the morality or the economics of it. What you don't know is the Southern man. He's charming and gallant and aristocratic. But he will have his flesh, dark meat or white. And as things stand now he can have both."

Harrah had never looked at slavery in quite this light. His own discussions, or at least the discussions he had heard, had always dealt with economics or questions of racial superiority or arguments for states' rights. He had never heard objections so nakedly put. This was a new voice and a new insight for him. He wondered whether Madeleine alone, being the sensitive woman she was, held this view or whether it was as passionately shared by other Southern women. Certainly these ladies could not be blind to what was going on in the darkened shacks that lay outside the pale of the big white mansions with their Greek columns.

"Please don't misunderstand," Madeleine said as an afterthought. "My husband never so much as touched a black woman. So I'm not mentioning this out of spite. But his father sired at least a dozen mulattoes. The imprint of his face is on all of them."

"You don't like the man?"

"He's dead now," she said softly. "But while he was alive I hated him."

"All the more reason not to go back."

She shook her head almost as a shudder.

"I have no choice in the matter."

Harrah threw his hands out.

"But," she said, "it would make things easier, infinitely easier, if you would come too."

Harrah stroked Madeleine just above the eyebrow, smoothing down a wrinkle.

"I'm not trying to be difficult, my dear. I want very much to be with you, to live where you live, to make your children mine. I've even tried to picture myself in Virginia—in some place other than the plantation. But I just can't bring myself to do it. I can't live the life of a Southerner, not in this day and age, not in these black times."

Madeleine moved away from him and began to dress herself.

"Then you don't want to marry me?"

"Of course, I'll marry you."

"But you won't live in Virginia."

"I don't know where we'll live. But we'll work something out. It may take a little time, that's all."

"I think it will take more than that, Nathaniel." She was now almost all put together, her yellow gown stretched wrinkle-free and her hair in place.

"I love you, Nathaniel. So I'll give you some time. I'll give you all of this summer and more. But one day you're going to have to make up your mind. And there isn't all that much room for compromise."

No, there wasn't, he admitted to himself. Not much room at all. And a Virginia Compromise, unlike Henry Clay's Missouri Compromise, looked very much in doubt.

1851

CAPE ISLAND INC.

INAUGURAL DINNER

In March the state legislature incorporated the City of Cape Island. In anticipation of this singular event Harrah had written Madeleine Culpepper, asking her to come to the Inaugural Dinner, set for the first day of Spring, and to stay a few days "to enjoy this glorious time of season that only the year-round residents get to know." He had put his best hand to the letter, hoping the extra effort would bring the desired result.

It did. But Madeleine's steamboat arrived a few hours late. Choppy waters and gray skies perversely delayed the boat's landing. Madeleine had barely time to freshen up when Harrah whisked her into a carriage and drove to Congress Hall.

"Do you always treat a lady so cavalierly?"

"I'm sorry, my love. But I didn't want to be late."

"Don't you have time to kiss me?"

"I did kiss you—at the Landing."

"You've done better than that in the past."

He pulled the carriage over and kissed her full on the lips, squeezing the breath out of her.

"Is that better?"

"It'll do—for now."

At the dinner Harrah was seated with some other, newer hotel owners. After ten years he had not yet been at Cape Island long enough to be included with the older, established families who had tables to themselves. In the past this type of discrimination irritated Harrah who by now regarded himself as much a Cape Islander as anyone else. He even made it a point to show his annoyance by excessively tipping his hat to the "natives" or by winking at their older daughters and then ignoring them when they showed an interest in him. But with Madeleine present he could not care less.

After a dinner of chicken and candied yams, the new, the first Mayor, Isaac Miller Church, was introduced to the gathering, although he needed no introduction. Since 1848 he had been Pastor of Cape Island Baptist Church, having served a few years as a missionary before taking the post. And Harrah suspected Church looked upon his new job as a mission, too. As the pastor, dabbing his mouth with a linen cloth, rose to speak in his new temporal capacity, Harrah realized how similar politicians and clergymen were. Both saw themselves as ordained to lead. Both gave the impression they had the ear of the Lord when seeking counsel or advice. Both came down heavy-handed when they encountered honest criticism, often branding it as heresy. And both made their share of clerical errors. Nevertheless, Harrah regarded the auspicious occasion with hope, promise, and buoyancy, and was anxious that Madeleine share his enthusiasm.

"His wife's a Cape May girl," whispered Harrah, leaning over to talk to Madeleine. "He looks older, but he's only thirty-seven."

"You're a fund of information tonight," she teased. "Most nights I can't get a word out of you."

"Well, this is a special occasion. It's a new era for Cape Island. We're a city now, not a town on a creek."

"You forget the ocean. It's the ocean makes Cape Island. Not its incorporation into a borough or city."

"How right you are, my clever one. But still, there will be changes."

When he looked up, Mayor Church had begun his speech, an updated and somewhat revised version of his inaugural address.

"It is now some fifty years since Ellis Hughes placed an advertisement in the *Philadelphia Daily Aurora*, informing the public of his inn (later called The Atlantic) and extolling the virtues of the sea and of Cape Island as a bathing resort. In those days Cape Island, indeed all of Cape May, was a sparsely populated place (boasting only a few cottages and some small farms) where fishing and agriculture were the main occupations.

"It was only thirty-five years ago that Tommy Hughes built a barn of a house, three stories high, that people called 'Tommy's Folly.' But Tommy's Folly, which was named Congress Hall in '28, became the forerunner of other large houses to come. It was also some thirty-five years ago that the Steamboat and Packet Line began regular runs from Philadelphia and New-Castle to Cape May.

"It was only nineteen years ago that Richard Smith Ludlam built the Mansion House, the second large hostelry erected on Cape Island, with music provided by Hazzard's band. And it was only eight years ago that the first church was built on Cape Island. We now have several houses of worship, including a Methodist, a Presbyterian, a Baptist, and a Roman Catholic Church.

"To say that we have progressed in fifty years from a small fishing and farming village to the premier resort of the Atlantic Coast is to say the obvious. But what of the next half century? Will we continue to see rapid growth and significant change? Although it is difficult to make forecasts, there is every reason to expect that the island's growth will continue at an accelerated pace, with more and more visitors from North and South flocking to our shores.

"And in the history of Cape Island, 1851 will stand as a high-water mark. Not only because Cape Island was incorporated as a city but because it is prepared to do what it must to realize its destiny. This past twelvemonth we've seen the addition of numerous hotels and cottages. We've surveyed and laid out six new public streets. We've had our first council meeting—on March 15th at the schoolhouse. And we've seen an improvement in the regularity of steamboat runs to Cape Island and the addition of ocean-going steamboats stopping here on their way to Philadelphia or New York."

"Hardly an improvement," interjected Madeleine so that only Harrah could hear. "My trip was dreadful."

"But much remains to be done. The threat of fire is forever with us. You've heard of the conflagration in Philadelphia that destroyed 354 houses last year and an untold number of people? To ensure that this does not happen on Cape Island, I am advocating the inspection of all buildings by fire marshalls and the requisition of ten fire buckets for every landing in our great hotels and houses and the formation of a hook and ladder company."

"Do you have fire buckets on your landings?" asked Madeleine. Harrah nodded.

"Strange, I never noticed."

Mayor Church went on. "There has over the years been a serious concern abut indecent or improper behavior on the bathing grounds. Most of the big hotels have set their own rules and regulations. And I have no quarrel with these. But I am going to recommend that stringent measures be taken to preserve the decorum and dignity of Cape Island where unacceptable behavior has become the mode.

"To our neighbors in what I shall call adjacent country, I should like to say this. We want your marketing, your fish, your hay, your wood, your labor, your teams, your vehicles. But it would add much to the respectability and comfort of Cape Island if no horses, cattle, sheep, goats or swine were allowed to roam at large within city limits."

"And I always thought that was part of Cape Island's charm," remarked Madeleine.

"You're full of mischief tonight," observed Harrah.

"I don't like speeches—especially on my first night here."

"He'll be done soon," said Harrah. "Then we'll return to Trescott House."

"I'd rather we stopped at your cottage."

"We'll have lunch there tomorrow," said Harrah, taking her hand.

"I don't know if I can wait till tomorrow."

Mayor Church at last reached the climax of his speech, "I don't know what the next few years will bring for Cape Island. The Good

Yevish

Lord in his Wisdom has not given us the power to look into the future. Yet I trust we shall not be deterred from launching Cape Island City, though the channel be narrow, shallow, and difficult to navigate. For if we do our duty, if we apply ourselves to the tasks at hand, if we avoid the sandbars and shoals that may impede us, I think we shall float out to good sailing."

"Well, what do you think?" asked Harrah.

"I've had enough floating for one day. Suddenly I'm exhausted. Let's go home." She squeezed his hand. "But tomorrow the cottage."

AT ODDS

Opening the cottage door was easier than in the past, for Harrah had hired young Henry Sawyer to fix it, along with the rest of the doors in the house.

When Harrah pushed back the shutters, letting the light in, and Madeleine saw the living room, she exclaimed, "How nicely you fixed it up! Where did you get such a grand couch?"

"From the Delaware House. James Mecray gave it to me when he enlarged the place. Used to sit in the parlor."

"And the oil lamp?"

"From the Columbia House across the street. The table, too."

"So, you've become the scavenger of Cape Island?"

"No, the savior of its fine furniture. Furniture that has been sacrificed to the fickleness of fashion."

She was amused but would not remove her cape. Harrah bent down to get the fire started.

"Besides, these things were given to me free of cost. All I had to do was move them here."

"I didn't realize you were so enterprising."

"It's the Yankee in me. Don't let the South Jersey exterior fool you."

"I meant it as a compliment. But now that you said Yankee—"

"Be grateful I'm a Yankee. I'm not so sure those from the Sunny South can start a fire."

"You don't seem to be doing so well yourself."

"It's lit." He stood back to watch the struggling flames catch on.

Madeleine sank back on the couch and Harrah took his place beside her.

"This is probably the only house we'll ever have." she said, leaning against his arm both for warmth and affection.

"I wouldn't say that."

"You know it's true. Not that I'm complaining about the cottage. At least it's 'ours'—in a very special sense. You will never live in Virginia, will you?"

He could not give her the answer she wanted.

"And I can't live on Cape Island—at least until the children are grown. By then I'll be too old and you won't want me."

"Why do you talk that way, Madeleine?"

She let him draw her closer but she did not take her eyes from the fire.

"I've given it a lot of thought this winter. I don't see how we will ever resolve this thing.''

She lay in his arms, becoming sleepy as the fire grew warmer.
''But I do love you.'' she murmured. ''I couldn't wait till that steamboat reached the landing.''

''Nor could I.''

He embraced her and she responded warmly to his touch, to
his kisses. He removed her cape and began to loosen her clothing.
She was half undressed, with one breast free, when she pulled away
from him.

''No, I'm not going to make love to you this time!''

Harrah was stunned by the sudden change of mood.

''Oh, I want to. I've hungered for you all winter. But if there's
going to be no life together, there's going to be no lovemaking
either.''

Harrah did not try to dissuade her—at least not yet. He knew
it would serve no purpose while Madeleine was in this turn of mind.
He went about tending the fire.

''You don't understand, do you?''

''I understand, Madeleine.''

''Then why don't you say something?''

''What do you want me to say? That I love you? You know
that already. That I'm not going to Virginia. We've been over that,
too. The fact is, I can't leave Cape Island. The only income I have
is derived from Trescott House. I've not invested in land or another
business or in the railroads. On Cape Island I'm a free man.''

''And in Virginia? Will you feel as much a slave as the Negroes?''

''More so, Madeleine. As a master I'd be a slave to an institution I deplore. At least your fellow Southerners give the impression of believing in it. I couldn't.''

''I don't believe in it any more than you do, Nathaniel. I don't
believe in the slavery called marriage either. Where women are
completely dominated by men. Still, I married. And would marry
again. One must be realistic.''

''Maybe I'm an incurable romantic.''

He drew her towards him once more.

''I love you, Madeleine. I want you as my wife. I want you
so badly I'd sell Trescott House tomorrow. But I won't live in
Virginia. Or any other of the slave states. I'd give up a lot of things
for you. But not my integrity as a man. What's being done to these
poor people is so cowardly, so self-serving that I couldn't look at
myself if I were a party to it. As far as I'm concerned, a man who
holds slaves has forfeited his membership in the human race.''

Madeleine pulled away from him again.

''So that's where it stands, does it?''

"Pretty much," said Harrah.

She turned her back to him and looked out the window.

"I'm going to allow you to make love to me, after all," she said quietly, almost as an epitaph. "I've waited too long to deny myself. Or you. It may be the last intimate moment we shall have together. And I want to remember it. Because, Nathaniel, when I leave Friday, if I leave uncommitted, I won't be back this year. Make no mistake about it. I mean what I say."

Madeleine Culpepper twisted her lovely neck and turned her eyes upon him. Then, stretching out her arms, she opened herself to him.

LEAVE-TAKING

If Harrah was delighted with Madeleine's presence on Cape Island, Willacassa was positively enchanted.

"Oh, I do love the gift you brought me. The bracelet is so beautiful, I can't wait to wear it."

"Wear it now, Willacassa."

"In these things? No, I'll wait for a special occasion."

Madeleine smiled warmly upon her.

"Even if you didn't bring me anything, I'd be thrilled at the idea of your being here. Do you know? It's silly for me to say it. But when you stand against the sky it's as if the Lord made Cape Island just for you. To provide the kind of seascape you deserve. You are so very beautiful, Mrs. Culpepper."

"Why, thank you, Willacassa. You are too kind and generous. I wish I were half as beautiful as you say. But I will accept the compliment anyway. Speaking of beauty, you have grown into a most attractive young lady. I can hardly believe how lovely and mature you have become—in less than a year."

"I'm filling out if that's what you mean. But that can be a problem, too."

"A problem?"

"The boys—or young men, as you would call them—have noticed too. And I would just as soon they didn't."

"Pay them no mind."

"I don't. And it infuriates them. But I don't really care. A woman has to face these things. And still be a lady for all that."

"How wise you are, Willacassa. It's a joy to talk with you."

"Oh, Mrs. Culpepper, please stay until summer. I would so love to have you about the house. Without you we're just three strangers. Four, counting Rachel. When you're here, we're a family."

"I'd love to stay, my dear. But the truth is, I have to cut short my visit."

"But why?"

Willacassa expressed such deep concern that despite her customary reserve Mrs. Culpepper felt she owed her an explanation. Though she did not want to be explicit about her relationship with Harrah, the words came tumbling out.

"I had hoped to be your father's wife, Willacassa. Yes, and your mother. I had hoped that the differences between us, between North and South, were not sufficient to cause a rift between us. But I see I was wrong. It's not his fault, this problem of ours. And it isn't mine. But I must go back to Virginia empty-handed and pray that one day things will be different.''

''Oh, no. Don't say that. Please don't go. I want you and my father to be together. You are so right together. And—''

''And what, dear?''

''And I want you, too. I love the person you are, Mrs. Culpepper. I don't want you to leave us.''

Madeleine Culpepper took Willacassa in her arms and tried to comfort the girl. She fully understood what Willacassa meant, having been a thirteen-year-old herself once and having had her own idols at that age. And she wanted to be able to say that she was staying. But she could not.

''I'll always remember how kind you were to me, Willacassa. It is more than I had a right to expect. I know that you will make us all proud one day. And yourself too.'' Madeleine Culpepper realized that she was not saying what she had wanted to say. But the words filled a gap and as surrogates for her real feelings did reasonably well.

When she was gone, Willacassa searched out her father. She was in an emotional spillover.

''What did you say to Mrs. Culpepper?''

''What do you mean?'' asked Harrah.

''She's leaving, you know!''

''Yes, I know.''

Harrah tried to move out of storm range.

''Don't let her go, Daddy. She loves you!''

''You don't understand, Willacassa.''

''Marry her. And then we can be together always.''

''She has her life, her responsibilities, and I have mine. They're not compatible.''

''Oh, piffle! You're just too proud to go to Virginia. Don't stay on Cape Island on my account. Or Jonathan's.''

''It's not that simple, Willacassa.''

''Simple! Complex! What difference does it make? You're letting the loveliest woman I've ever known get away from you!''

Harrah turned his back on his daughter. He did this not to silence her but to stem the emotions that were creeping up on his heart, threatening to stifle him. With both hands he seized the back of a chair and squeezed hard.

''Jonathan!'' called Willacassa. ''Jonathan, get the carriage out. I'm going to the Landing.''

She turned to Harrah. ''You won't stop me, Father. Will you?''

Harrah shook his head.

Willacassa tore along the road leading to the Landing, her hair flying in the wind. Henry Sawyer had taught her to drive a carriage, and she whipped her horse with all the abandon he was capable of when he had put away a beer or two. Even the rude bumps could not halt her breakneck pace.

When she drew near to the Landing, she could see Madeleine Culpepper, a forlorn figure against a dull sky, sitting on her trunk, watching as her boat tied in at the temporary dock.

"Madeleine!" she called. "Mrs. Culpepper! I've come to take you back!"

Madeleine Culpepper stood up and turned round to see a hysterical angel of mercy draw to a vigorous stop.

"Why, Willacassa! Wherever did you learn to drive so hard?"

Willacassa jumped down and threw her arms around the woman.

"You must promise me you won't go!"

"But I must. My children are waiting."

Willacassa backed off. "Then bring them here. We all want you. We need you on Cape Island!"

Madeleine Culpepper glanced at the seaman who was motioning her aboard. "I wish I could, dear. But it's not as simple as that. Didn't your father explain things to you?"

"Lady," interrupted the seaman, "the boat must leave. We're late already."

"I know. I know. Just a moment longer."

As she said this, she was seized by Willacassa in a frantic embrace. For Madeleine there was now a collapse of all composure. The Virginia beauty squeezed her eyes closed and wrinkled her lower lip to stifle her sobs.

"Oh, Willacassa! Willacassa, my dear! How difficult you are making this for me."

"Don't go," sobbed Willacassa. "Don't you love me?"

"Of course I love you, my child. My heart aches with love."

"And father?"

"And I love your father too."

"Then how can you leave?"

"Because we're creatures of different worlds. That's how."

"That's no reason, Madeleine. I can call you Madeleine, can't I?"

"Of course, my sweet."

"If you love each other—"

"It's not enough to love someone." By now Madeleine Culpepper was in control of herself once more. "Our situations are at odds with one another. It's no one's fault, really. Just the Lord's way of seeing we're not too happy."

"Not too happy?"

"Because of all the pain we've inflicted. You'll know better what I mean when you're older."

Willacassa pushed away from her.

"No, I'll never forgive you for leaving!"

"But I can't stay," intoned Madeleine.

"You're both too stubborn in this thing. Someone must bend. There's no bending Father, you know."

"If it were solely up to me, I'd stay. But I have my children, Willacassa. And I must think of them. I must put their lives before mine. Just as your father puts you and your welfare first."

Willacassa shook her head. The young lady that people were beginning to see in her all but disappeared.

Madeleine Culpepper put her hand on the girl's shoulder. "The last thing I want is to leave you, Willacassa. You must believe that. Now kiss me goodbye," she said softly, beseechingly. "And let me go. Or I'll die on this Landing of a broken heart."

Willacassa looked sternly at her, then softened.

"I'll miss you, Madeleine." She kissed the woman, then released her. 'But I'll never forgive you."

"You will—one day."

Madeleine picked up her parasol and handbag and with the help of a seaman stepped up the gangplank. When she reached the rail of the steamboat, she looked for Willacassa and waved sadly until the boat pulled free.

1852

WIDOWER'S WALK

OF TIME AND THE OFFSPRING

It was not the clock or the pendulum that measured time for Harrah, though he had clocks enough in Trescott House. Nor was it the calendar in The Old Farmer's Almanack, a difficult thing to keep track of unless one remembered to cross out the squared days that had already passed. No, it was the ebb and flow of the ocean as he saw it from his window, the back and forth throbbing of the waves, a watery heart muscle eternally pumping, pushing forward, then surging back, rushing towards the beach, then quietly retreating, leaving only wet sand to show where it had been, that measured the passage of time. And later as he walked the strand it was the pounding, roaring, swirling, lapping, whispering flow of water that murmured the months away, turning time into eddies and endless swirls.

For it was almost a year since he had seen Madeleine Culpepper. And in another month he would know whether he was to see her at all in the Summer of 1852.

Soon Harrah was too busy to measure time by clock or by ocean. Summer visitors were disembarking daily at the Landing and hailing carriages or farmer's wagons to take them to the heart of Cape Island. Soon Harrah had a houseful of guests who had to be provided with bed and board, the finest wines, the latest magazines and newspapers, comfortable chairs, soft cushions, decks of cards, and oval green-cloth card tables.

Willacassa was a big help these days. Past fourteen, she was taller than average, strong, with a rich growth of blonde hair tied back, and a developing bosom. In the mornings she did the day's shopping, and upon her return helped Rachel in the kitchen, slicing vegetables, preparing salads, and brewing Trescott House's aromatic coffee. At noon she filled in as waitress, setting tables, serving the viands and comestibles, and carrying off the dishes. She had another local girl to help her and between the two of them twenty-five guests were served.

Jonathan, now seventeen, had his chores too, everyone of which he hated and railed against *sotto voce*.

"Why the devil do I have to make beds! I'm not a slave! If the chambermaid is sick, hire another one."

Back in the kitchen where he had to clean fish scales from the would-be evening entree, he offered these comments.

"Why can't we serve meat or chicken? I hate scraping scales. I smell of fish all day."

"We're having chicken tomorrow," announced Rachel. "You

can pluck the feathers.''

"Why can't we buy them already plucked?''

"We do," said Willacassa. "Rachel was only teasing. What an unpleasant fellow you can be at times!''

"It's just that I'd rather be playing quoits with the railroad men or having my shots at the archery tents.''

It was not until Jonathan shook the chain of his chores each day that he tasted the liberated air of Cape Island's carnival life. Jonathan had early acquainted himself with the billiard parlors, the archery tents, the ten-pin alleys, and the taverns of the resort, none of which flourished in the off-season except for the taverns. His interest in girls, too, was keen, particularly the perky ones. And he loved to dance. But this took second place to his other pleasures.

Jonathan took advantage of the beach only so far as it provided a setting for a game of quoits or for carriage racing on the strand. As Harrah did not permit him to race his one carriage on the strand, the youth was limited to wagering on those that did. What money he earned he quickly spent on a slow pair of wheels or a one-lunged horse. It was said by Willacassa that he had a hole in his pocket almost as big as the hole in his head.

None of this phased Jonathan. He assured Willacassa and anyone else who would listen that he would make his fortune before he was twenty and that he would then build a guest house twice the size of Congress Hall.

"Then you'll have that many more beds to make," teased Willacassa.

"And fish to scale," added Rachel.

"Your son shoots a mean arrow." remarked Henry Sawyer one day as he repaired the stall Harrah's horse had kicked out.

"What do you mean?"

"Jonathan may not be proficient with hammer and nail," said Sawyer lying on his back and working overhead. "But his dexterity at quoits and archery leave him few competitors."

"He's that good?" asked Harrah with a trace of pride.

"That good."

"But what will it do for him? You can't make a livelihood as a South Jersey Robin Hood."

"I'm not so sure. Half the time Jonathan wagers on himself. And more than half the time wins."

Harrah stepped into the tack room to rearrange his leather. "I hear that he's no stranger to the Kursaal."

"He's there every evening," observed Sawyer. "If not dancing, listening to the music, and watching the theatrical performances."

"Do you see any harm in it?"

"What harm can there be, Mr. Harrah?"

"Too much of a good thing."

"Oh, it's that all right. But it's where the pretty girls are. Properly chaperoned, of course." Sawyer laughed at the idea, and Harrah wondered if he were laughing at Jonathan too.

"The boy worries me, Henry. I've not been father or mother to him. Least not when he needs it most—in summer."

"None of us have time for our own in summer, Mr. Harrah. If you're going to make a dollar, it's summer or never."

"Maybe I'm in the wrong business."

"Living in a resort town, you mean?"

Harrah nodded.

"Hasn't hurt your daughter any. A regular young lady, that girl. Shops, waits tables, pays the bills. Hope my own kids turn out as well when the time comes."

Sawyer drove in his last nail and finally slid out from under the stall.

"That ought to do it, Mr. Harrah. What got your horse so riled up?"

"Jonathan, I think. He took him for a ride and didn't dry him off properly. Didn't feed him either. The horse likes a little reward now and then."

"Don't we all?"

"Will you be able to work on the steps too?"

"Maybe a little later in the week."

"I don't want anyone to break his neck."

"Did they pay their bills?"

"In advance," said Harrah. "House policy."

"Then I wouldn't worry about it," laughed Sawyer, brushing himself off and throwing his jacket on his muscular shoulders.

"And I needn't worry about Jonathan?"

"I'm not saying that, Mr. Harrah. He needs some toughening up. Some limit on his leisure time. Maybe only once or twice a week at the Kursaal. But he's not without skills of his own. Just got to put them to good use."

"Thank you, Henry. Sounds like good advice."

That evening he found Jonathan in an ugly mood.

"What hornet's nest did you sit on?" asked Harrah.

"It's that Miller boy. He owes me money and won't pay up."

"What for?"

"I said I could get him some magic tricks. And I did. Brought in by Captain Seymour from Philadelphia. But he doesn't want to pay the price."

"Maybe you're asking too much."

"I took the trouble to get it, didn't I?"

"Actually it was Captain Seymour got it."

"But I made all the arrangements. That's worth something, isn't it? I'll be damned if I extend myself the next time."

With Jonathan, Harrah realized, it was sometimes better to let matters rest. Youth had to have its day, even if it was predictably topsy-turvy. Harrah knew that he should spend more time with his son, that these were the developing, the critical years in their relationship. And he regularly resolved to find a greater part of the day to be with the boy, to share in his pleasures and summer pastimes. But in July and August this was easier promised than done. During the off-season Jonathan had his school books and papers to keep him out of mischief. But, though Harrah had broached the subject, Jonathan never indicated an interest in pursuing his education at Princeton College or Rutgers. This meant leaving home, and though the youth talked about being independent he secretly feared cutting out on his own.

"If I ever leave Cape Island," he would say, "it won't be to sit in some stuffy classroom. It'll be to make my fortune. Either by gambling or investing in the railroads."

"That's pretty much the same thing," chided Harrah, quietly condemning himself for ever having introduced gambling at Trescott House.

"It's all talk," criticized Willacassa. "He's afraid to leave the island. Might get kidnapped by a highwayman or lost at sea."

"I'd leave tomorrow if the right proposition presented itself."

"A proposition doesn't present itself, Jonathan. You've got to go after it," insisted Willacassa. "What makes you think one of these rich summer visitors is going to make you of all people an offer? How tiresome you get with all your jabber!"

"You'll see, Willacassa. You'll see whether it's jabber or a promise kept."

But try as he would Harrah found little or no time for his son. The incessant demands of Trescott House, the fuss with accommodations, the fastidious demands of the cuisine, the petulance of house guests who after all were paying a tidy sum for their short stay and wanted their money's worth—all this kept Harrah from taking the day, a half-day, even an hour to accompany Jonathan on his peripatetic journey through Cape Island's diverse amusement places.

They did manage to fire pistols at a shooting gallery. And Jonathan was astonished by his father's accuracy in marksmanship, checking the target not once but three times to count the bull's eyes.

"Where'd you learn to shoot like that?"

"During my training as a treasury agent."

"You were a treasury agent?" Jonathan's admiration was unabashed.

Harrah nodded but quickly changed the subject.

HARLEY COLEPAUGH

In the absence of Madeleine Culpepper, in the painful recognition that not only his happiness but his ability to function was at stake while she was away from Cape Island, Harrah attempted to reevaluate his feelings about himself, the South, and the whole slavery question. He wanted to be sure of his ground, especially if taking that ground was to give him such undiluted pain. If he could in some way extricate himself from his philosophical position, his supposed convictions, he could perhaps find a way out of his dilemma with Madeline Culpepper. And in finding a way out, he could perhaps find an honorable solution that would both satisfy his conscience and his unfulfilled desire for a life with her.

But struggle as he would, he could not put together a rationale, a patchquilt of reconciled contradictions, that he could sleep with at night and could look at without flinching in the honest light of morning. The truth, unlike the Biblical dictum, did not set him free. It taunted him, it mocked him, it threw up a barricade around him and with nagging doubts barred him at every turn from taking any action that would relieve the burden on his heart. He wanted to be a hyprocrite but could not. It remained for Harley Colepaugh to set him straight.

The Fugitive Slave Act of 1850 had made little impact in the North. Harrah had only to read the newspapers to see that it was honored more in the breach than in its enforcement. Occasionally an advertisement appeared in the New Jersey State Gazette out of Trenton or one of the Philadelphia newspapers offering a reward for the return of some unfortunte wretch who had taken it upon himself to ensure his freedom.

To Harrah's dismay and personal feeling of shame, New Jersey, which as a border state had its share of runaways, stood alone among the Northern states in supporting enforcers of the act and in offering no opposition to the seizure and return of fugitive slaves.

One of the few church groups that railed against the Fugitive Slave Law was the Presbyterians who declared it "unconstitutional, subversive of morality, oppressive to enlightened freedom." They declared they would sooner suffer the penitentiary than cooperate with such an unjust law.

Harley Colepaugh was a Presbyterian, but his branch of the church lay in the South. And, as he put it, Presbyterian meant a different thing in North Carolina than it did in New Jersey or Pennsylvania. Even if it didn't, Harley Colepaugh made sure at every turn of a playing card and every swig of a pint of bourbon he furtively carried in his coat pocket that he would get his lumps in as far as runaway slaves were concerned. It wasn't that Harley Colepaugh was any more hog-wallowed than his compatriots from the South or any more insensitive to Northern feelings. He was just less inhibited than his fellow cardplayers in his manner of speaking. And good hand or bad, he felt it his duty to render his pronouncements. But when he held a good hand he did manage to edge his comments with humor, not so much to take the sting out of his remarks as the make them more memorable.

"You should have seen that darkie run. I wasn't goin' to hit him with that cat-o'-nine. I was just goin' to wave it over his head. But he ran so fast he left his sweat taggin' behind him. Got picked up though before he crossed the border. And then I laid it into him."

"Bet that whip was made in Newark," ventured his opposite at the card table. "We manufacture all kinds of leather goods in Newark. And I'd say better than two-thirds of our shoes, clothes, and carriage harnesses are marketed below the Mason-Dixon line. You might say the South walks on Jersey leather."

"Maybe the South walks. But our niggers shuffle and they do that foot bare. I suspect that's why the abolitionists want to let 'em loose—so they can sell them shoes made up North!"

"Come on, Harley. Play your hand."

"How do two aces look to you, my friend?" Colepaugh spread his cards.

"Not half as good as two pairs. Your deal."

"Holy tarnation!" screamed Colepaugh. His choleric outburst lit a fire under the brassy sheen of his hair. "How the hell did you get two pair!"

Springing up, he deliberately capsized the table, scattering coins and cards, ashtrays and glasses. His companions gaped in astonishment, backing away to avoid being spilled on.

Harrah was generally pretty tolerant of his guests. A carpet spill, a table nick, a broken dish—this was part of the burden of being a hotelkeeper. One made the best of small mishaps and charged it to petty cash.

But Harrah disapproved of Harley Colepaugh with a passion. And Colepaugh's was an act of commission, not omission. Having witnessed the upheaval, Harrah was not going to overlook it.

"Pick it up," he said to Colepaugh, as close to anger as he had ever gotten with one of his patrons.

"You talking to me, Sir!"

"I said, 'Pick it up!'"

"I ain't no nigger."

"If you mean Negro, no Negro knocked it over," said Harrah. "Pick it up or get the hell out of here!"

"I've paid in full. And I'm stayin'"

"Your money'll be refunded. Except for damages which I'll deduct."

"Come on, Harley. We'll help you," said one of his compatriots.

"Hell, no! I ain't pickin' nothin' up." He started for another table, to upset that one too. But the look on Harrah's face put a stop to that idea.

When the other cardplayers righted the table he had upended and began tidying things up, Colepaugh did help with the glasses.

"Got to pour myself a drink," he said, wiping one of them on his sleeve. Then he filled the glass with bourbon and downed it at a shot.

This time the stimulant had a quieting effect on him. And in a few moments the gameroom returned to normal.

"I want to apologize for my compatriot," said Mr. Arrows later on. "He's had too much to drink. We Southerners may have our own prejudices but we are first and foremost gentlemen."

"I have no complaint about you, Mr. Arrows. Or any of the other guests here. Most of the Southerners I've met comport themselves very well. Harley Colepaugh is the exception, drunk or sober. He does his cause no honor."

"There are some of us," said Mr. Arrows, "who feel the cause is deserving of no honor. Not everyone in the South approves of slavery, Sir."

"I know that, Mr. Arrows. On this issue at least, North and South aren't all that clear-cut."

"Perhaps one day the institution will die a natural death. I, for one, hope so."

"A natural death? I don't think we have that much time, Mr. Arrows. Not while there are Harley Colepaughs around. And Fugitive Slave Laws to shore him up."

OF TIME AND THE HIRED MAN

Harrah had to get in a supply of corn and sweet potatoes, not to mention ham, flour, and corned beef. His weekly delivery to Trescott House had not been made, and Harrah had no word from his farmer as to the cause of the delay. As the prices at the stores were fearfully high, typical of mid-summer, and everyone seemed to be low in stock, he ruled that option out. Hiring a wagon in case he had to cart the produce himself, he prepared for the worst.

"Husband's been laid up. Still running a high fever, Mr. Harrah. The crops are there, but they've got to be picked. We've done no slaughtering for five days now. Can't get anyone to help. They're all down at the Landing, carting passengers and baggage."

"What am I to do? I'd pick the stuff myself but I've got to meet people at the Landing, too."

"Don't know what to tell you," said the farmer's wife whose strained face suggested she might be down with fever herself soon.

Harrah drove up and down the back roads, looking for someone to hire on the spot. When he had all but given up, he saw a colored man in a wide-brimmed hat, walking head down along a wagon path.

"Hello there!" he called. "How'd you like to make a dollar or two?"

Either the man did not hear him or he pretended not to understand.

Harrah drove up alongside the man, then leaned across the driver's seat to address him.

"Would you like to make some money? There's some corn and potatoes need picking. I'll give you a day's pay for half a day's work."

The man, it appeared, could use the money. His trousers were worn through and stained. His jacket was torn at the pockets and had sticks of straw clinging to it. His shoes had no backs and were broken open at the toes. Still the man kept walking. If anything, he kept his head closer to the ground, not once looking away.

Harrah brought his wagon to a stop and leaped down upon the grassy path. Taking the man by the shoulder, he said, "Look here, my good friend. I'm offering you some work—"

But his contact had the opposite effect from the one intended.

"You don't want me. I gots to keep movin'. I cain't help you no' anyone else."

There was such pale fear in the man's eyes, shaded as they were by the broad-brimmed hat, that Harrah drew back.

"Jus' forget you see me," begged the colored man, stepping up his pace. "Jus' forget I ever walk on this path." When he had pushed some distance between them, he turned to look at Harrah as if to fathom his reason for being at that particular spot in the world at that particular time, a spot the black man had thought he had all to himself. Then, moving almost backwards, he edged towards the woods. Once he had reached the trees, he turned and darted among them, quickly disappearing.

It was then, and only then, that Harrah recalled the newspaper advertisements he had seen, those little boxes of black print that looked like death notices, offering small rewards and pitifully brief descriptions. But what the print had limned so matter-of-factly in the abstract had suddenly become real and painfully alive. And unforgivable.

WHY CAPE ISLAND?

It had been a long year for Harrah. And as the months drifted slowly into the sand and the ocean pulsed its waves against the dune barriers to the east, Harrah asked himself, Why Cape Island? It was enough at first that the town seemed a glorious setting for his children. A lofty lighthouse, fleecy clouds, a distant horizon, and an endless stretch of pure white sand to cavort on—what more could a man ask for his offspring? Besides, it had none of the closeness of the Pines, the shady, speckled, green and outcast world that Cassandra, their mother, had loved so much and Harrah had found so confining.

Later Cape Island took on a new meaning. Like the hotel flags that caught the stiff breezes, it spelled freedom to Harrah. Though south of the Mason-Dixon line and a favorite vacation spot for Southerners, it was far enough removed from the stench of slavery and the stunted mentality that could justify such an abomination that it left the manacles of its master-slave society at Cape Island's very doorstep.

In that light Harrah could understand why he could not live in Virginia—or anywhere in the South for that matter. (And why he had to give up any idea of marrying Madeleine Culpepper.) But he could not quite understand why he chose year after year to remain on Cape Island. Other than its special summer charm and its unique geographic position at the meeting place of the Atlantic Ocean and Delaware Bay, it had no moral or strategic importance. As time passed, Harrah had become more and more isolated there, more and more a solitary figure. It seemed almost as if he were waiting for something to happen. But nothing of moment could occur at a summer place. People at play did not shape world events.

"You're looking straight at me, but you don't recognize me. Have I changed so much?"

The truth was, Harrah had not seen her. She existed only in his mind's eye. Invention filled the void caused by her absence.

"Madeleine, you haven't changed at all. I guess I was preoccupied."

"How are you, Nathaniel? You don't look as happy as one might expect the proprietor of Trescott House to be."

"I'm happy enough, Madeleine. It's just that Cape Island seems so remote at times."

"Then why do you confine yourself to Cape Island? I'm sure your children would do just as well elsewhere."

"I was just thinking about that," said Harrah.

"About what?"

"Why Cape Island has such a strong hold on me."

"And what conclusion did you come to?"

"I came to no conclusion. All I know is that something about the place beckons me. I suppose it has something to do with miscegenation."

"The mixing of the races?" Madeleine was surprised.

"Yes, the coming together of North and South. True, the Northerners who come here are no more representative of the abolitionists and their ilk than the plantation aristocrats are of the ordinary Southerner. And in some ways the sympathies of the North have the same coloration as those from the South. At least they seem to be in agreement about the superiority of the white man over the black. But the important thing is they come. On this little island—nowhere else in this great nation of ours—two notions of society come together, two separate worlds really, that have no more in common than an English parliamentarian and an absolute autocratic Russian."

"But they must have *something* in common," said Madeleine.

"The only thing they have is envy. Envy on the part of the Northern entrepreneurs of the bondaged work force always at hand—no one running off to California or to homestead in Kansas or Indiana. Envy on the part of the plantation owners that money could be made in commerce or manufacture, made plentifully and easily, without having to endure the criticism coming from non-slaveholders."

"Oh, Nathaniel, you're getting too morbid. You've lived in isolation too long. I'm not suggesting you marry again. But there's something to be said for a good companion."

"I'll bear that in mind."

And so the summer passed, and fall chilled the waters off Cape Island. The year-rounders began storing up cordwood, and oil for the lamps of winter. And Harrah had the gnawing feeling that portentous events would add to the unrest he was already experiencing. He began to pace up and down his room as though it were a widow's walk. But he saw nothing on the ocean's bosom or the distant horizon that gave him any hope or sense of comfort. No ship was carrying Madeleine Culpepper back to Cape Island. No steamboat was pointing him towards Virginia.

Then another box of black print, a large one this time, to announce the death of Daniel Webster. Henry Clay—Harry of the West, the Great Compromiser—had died in June. How Cape Island had mourned the man! And now the silver-tongued giant from

Massachusetts was gone. Harrah could not remember when the great orator and statesman was not around, was not a figure to contend with on the national scene. The newspapers always carried one or more of his pontifical announcements. It did not matter that he nor Henry Clay never made it to the President's office. They were both giants. And now Webster too was stilled forever.

And with his silence there loomed an even greater threat to the peace of the nation, a nation that was moving inexorably towards war. Harrah had only to listen to his guests, both North and South, to see it coming. Nullification, States' Rights, Property Rights, Free Soil—these were only code words for the lighting of the tinderbox. Harrah himself did not see liberty and union as one and inseparable, as Webster had phrased it. There was union but no liberty for the Negro. And he felt the slavery question must be resolved—should have been resolved earlier—even if it meant pulling the Union apart.

And while Harrah had regarded Mr. Webster as a man of principle when he opposed the annexation of Texas and the war with Mexico, he did not regard him as a man of principle on the slavery question. Slavery was an evil, Webster had conceded; but he had always maintained that disunion was a greater evil. Well, Harrah was not so sure. If slavery was to be the glue that kept the nation together, he would just as soon see it break in two.

"Oh, Madeleine, why did you go back to Virginia! Why couldn't you see that there was no such thing as a compromise with chains?"

If as a kind of protest Madeleine had stayed on at Cape Island, he could somehow believe that the beginning of the end for slavery was at hand. But so long as she was in Virginia and gave by her presence a semblance of legitimacy to the bastard institution, the chains of bondage would invite infection—a festering, poisonous wound that only a massive bloodletting could cure. And bloodletting was the one thing he did not want.

1853

FIRST LOVE

ABOUT WILLACASSA

The Reverend Moses Williamson called on Harrah. For the past nineteen years he had been pastor of the Presbyterian Church at Cold Spring. A thorough scholar who had attended Jefferson College in Canonsburg, Pennsylvania, Dickinson College in Carlisle, the Theological Seminary at Princeton and who had studied Hebrew scriptures at Andover Theological Seminary, he had some time ago—at much expense and unflagging effort—erected an Academy at Cold Spring. It was at this academy, at Harrah's insistence, that Willacassa was now pursuing her education, her brother Jonathan having preceded her there.

A robust man who belied the story that he had come to Cape May for his health, Reverend Williamson had a shock of dark hair in the John C. Calhoun style, bushy eyebrows, a no-nonsense acquiline nose and a chin as firm and prominent as a Cape Island beach boulder.

"I've come about Willacassa," he said.

"Is she a discipline problem?"

"No worse or better than the boys."

Harrah was relieved. There was a strain in Willacassa that made conformity difficult for her.

The Reverend Williamson took the chair Harrah offered him. Harrah would have offered a brandy, too. But he could not remember whether the clergyman by reputation was an imbiber or a teetotaler.

"Did you know that your daughter has considerable talent as a sketch artist?"

"Talent, yes. Considerable talent, no."

Reverend Williamson pulled some drawings from his briefcase. "Do you recognize this?"

Harrah peered at the portrait. "Yourself?"

"A good likeness, no?"

"Quite good," admitted Harrah.

"And here's a sketch of the Academy with the children at recess. Notice how free and natural the forms are. I don't think she took half an hour to do it."

Harrah was impressed. He had always been aware of an artistic streak in Willacassa but he gave it no encouragement. An artist's life held little allure either at present or in retrospect, especially for a woman.

"There is talent in the family, I trust," suggested Reverend Williamson. "It is usually the way with these things."

"Her mother had a good eye. But the profession for a woman is difficult at best and in these parts a road's end."

"And you never put your hand to it?"

"I tried if for a while, but I had other things in mind."

"Then I take it," said the Reverend, "that you are not inclined to encourage her."

"Not really," acknowledged Harrah.

"Then what would you have her do?"

"Marry, and let her husband's profession be her own. If Willacassa stays half as attractive as she is now, she'll be quite fetching in a few years."

"I'm disappointed in you, Mr. Harrah. It's bad enough that you're not a member of my church, any church really." He smiled at his little jibe. "But you cling to notions that are fast becoming outdated. Women cannot forever live in the shadow of men. There are too many widows and spinsters abroad to find any solace in that. When a young woman has a special talent like your daughter—who will be a young woman soon—it is downright criminal to deny it."

"Perhaps you're right, Reverend Williamson. Let me give it some thought. In any case, I thank you for your interest. Is Willacassa doing well in other areas?"

"As well as can be expected. At least as well as the other girls. The boys, I'm afraid, are much more proficient in Mathematics."

He rose, tucking his briefcase under his arm.

"Good day, Mr. Harrah."

Having been told of Willacassa's talent, Harrah decided to find out for himself whether the girl had a career in the making. Certainly there was more opportunity on Cape Island for an artist to find portrait commissions than in the iron towns and taverns of the Pines.

"Here," he said, sitting down in his favorite chair. "I hear you've got a good eye for faces. Let me see what you can manage with this one, Willacassa."

"Yours?"

Harrah nodded.

"I couldn't do your likeness, Father."

"Why not?"

"You're too close to me. Too close to my heart. I couldn't be objective."

"Give it a try anyway."

"A profile?" she asked.

"No, a full face. Anyone can do a profile."

"All right."

She went upstairs to her room for a sketchpad and charcoal. When she came down again, Harrah could see she was high with excitement over the project.

"You'll need an easel, won't you?"

"No, the pad has a firm cover."

"Well, it will do for now. But I want you to get an easel."

"I'll order one from Philadelphia," promised the girl as she pulled up a chair and began putting charcoal to paper.

From her unusual way of looking at him, with a decided squint, Harrah could tell that she was working on his eyes first. Then she did some careful drawing—the nose, mouth, and chin no doubt. She would look at him as she worked, then at her paper. And with each rush of strokes she would lean back to compare the two, the paper and the man. Later she dabbed furiously and smeared the charcoal with her fingers, developing shadow and contour.

When with a flourish of the hand Willacassa indicated that she had finished, Harrah rose to study himself in the curious blend of line, shadow, and curve.

"A good likeness," he admitted, marvelling at her accuracy, "but a hurried one. Don't you want to get more detail into it?"

"It's not a portrait, Father. It's a sketch. I have no patience for full-blown portraits. And the truth is, I much prefer to do people in action. Shall I show you some of the bathing scenes I've done at the beach? Or the crowds at the pistol gallery?"

Harrah found all this a revelation.

"You mean you've put together a portfolio already?"

"A small collection of things."

"By all means, show them to me."

"I should warn you, they're only sketches. If you're looking for all kinds of detail, you won't find it here."

She ran upstairs again and in a few minutes returned with a sheaf of papers.

Harrah was impressed. The bathing scenes, awash with movement, had all the spontaneity and vigor of young people tumbling in the waves, and had just enough detail of bathing house and distant beach house, with flags flying, to complete the picture. The crowd scenes at the pistol gallery were so faithful to the push and crush of the moment and the fever of excitement that Harrah could almost smell the rancid puffs of gunpowder and the curling, twisting smoke of segars.

"Well, what do you think?" asked Willacassa.

"I think we have an artist on our hands," said Harrah with a mixture of pride and apprehension. For though the artist brought a pastel box of infinite possibilities with her, exotic chalks that tended to put flamboyant colors on the gray cast of everyday life, Harrah did not know if he was prepared for another go-round with this kind of heightened sensibility. One Cassandra—or at least her artistic temperament—was enough for a single lifetime. He did not know whether he had sufficient stamina left for another, even if this one was a daughter, not a lover and a wife.

MARCY CRITTENFIELD

One morning Willacassa told her father that a Miss Marcy Crittenfield was waiting to see him. Harrah indicated that he would meet Miss Crittenfield in the drawing room, but his daughter advised that the young lady wanted to see him in private.

When Willacassa ushered her into his office, Harrah saw a young woman of twenty-two or twenty-three, brightly dressed in colors he had come to associate with Baltimore rather than Philadelphia or New York. A slightly Southern accent, a coquettish swing of the body, and a brilliantly fetching smile convinced him he was right.

"Marcy Crittenfield, Mr. Harrah."

"No relation to Dr. Marcy?"

"I'm afraid not. I was Christened Marcia, but I much prefer Marcy."

"What can I do for you, Miss Crittenfield?"

The smile all but vanished.

"Hasn't Jonathan told you?"

"Told me what?"

"That we're planning to be married."

Harrah who had gotten up to receive the young lady now sank back in his chair.

"Married? But he's only eighteen."

"Well," said the young lady almost in a huff, "I'm little more than that. And Jonathan is quite mature for his age."

"Is he? Please sit down, Miss Crittenfield. You'll forgive me for being somewhat surprised. Jonathan made no mention of you or a forthcoming marriage."

"Well, I can assure you it's no figment of the imagination."

"Where are you staying, Miss Crittenfield?"

"At the Centre House."

"Have you known Jonathan long?"

"Since July Fourth. We met at the fireworks display at Congress Hall."

Though she was pretty enough, with fair skin except for a freckle or two on the cheeks and sparkling blue eyes framed by ringlets of black hair, Harrah was not at all impressed with the young lady. He found her overly dressed and too heavily powdered. And she lacked the maidenly modesty he looked for in young women.

"Don't you think Jonathan is too young to be married?"

"It was his idea, not mine."

"There is no great compulsion to marry on your part, is there?"

"What do you mean, Mr. Harrah?"

"As you've only met a couple of months ago, there seems no reason to rush headlong into marriage."

"As I've already said, it's Jonathan's idea, not mine."

"Then you would have no objection to a delay. Or even a lengthy engagement?"

"But I do object. We're very much in love. We're anxious to build a life together."

"I understand. But Jonathan is in no position to marry. He has no profession or trade. Nor is he preparing for one."

"Oh, my father will find something for him."

"Your father?"

"He's a railroad man from Camden."

"I see. And he knows Jonathan?"

"Very well. And he's most anxious that we marry."

"Miss Crittenfield, you will forgive me for being direct. But are you pregnant?"

"Pregnant? Heavens no!"

"Then what is the great hurry?"

"It's Jonathan. Why don't you speak to him before accusing me."

"I was not accusing you. I simply wanted to know where things stood."

Harrah surrendered his chair, hoping in this way to wind up the interview, but he was not quite through.

"Yes, I will speak to Jonathan. And maybe I'll find out what is going on. Meanwhile I suggest you mention this to no one you haven't spoken to already."

"Not mention it?"

"It might be awkward if there were no wedding after all."

"Then you don't approve of me?"

"I don't approve of eighteen-year-olds getting married, Miss Crittenfield. At least eighteen-year-old boys. It's nothing personal, I assure you."

Marcy Crittenfield flashed a brilliant smile once more.

"You speak to your son, then. After that, we'll have another chat. It was nice meeting you, Mr. Harrah."

When she was gone, Harrah stepped outside his office and wiggled his finger at Willacassa.

"Find Henry Sawyer for me, my sweet. I want to see him right away."

The young carpenter appeared in his workclothes, but his jaunty cap more than made up for a lack of grandeur.

"You want to speak to me, Mr. Harrah?"

"Yes, Henry. Do you have occasion to work at the Centre House?"

"As a matter of fact, I'll be going there this afternoon. Some floor boards need replacing on one of the verandahs. They've got a rowdy bunch over there. Knocked over a lamp during a party and started a small fire. Luckily it was put out. But one of these days we'll get a fire can't be stopped. And then we'll sizzle for blocks on end."

"I've a favor to ask of you, Henry."

"Anything, Mr. Harrah."

"It requires the utmost discretion."

"You have it."

"I want to inquire about a young lady called Marcy Crittenfield. I'm told she wants to marry my son."

"I'm not surprised," said Henry Sawyer.

"You're not?"

"No, I've seen Jonathan about town with a young lady on his arm. He appears smitten with her."

"She's older than Jonathan?"

"By a few years, I would say. I never mentioned it to you because I didn't think it was serious. He's run around with girls before."

"Well, it's serious this time. And I want to know more about the girl. Not that I'm going to let him marry at eighteen."

"I understand."

"Then you'll do it?"

"I'll do the best I can."

"Drop by after work and we'll have a drink together, Henry."

"Very good, Mr. Harrah. I'll see you then."

Harrah was not exactly pleased with what he was doing. But he saw no escape for it. If at his tender age Jonathan was involved with a girl, and the girl was at least four or five years older than his son, Harrah would have to know more about her—especially if there was a marriage in prospect. It seemed he did not know his son as well as he should. And Harrah berated himself for not being on closer terms with the boy. Had he spent more of his time with him, had they established a firmer bond, had the boy felt freer to talk with him about the things close to his heart, there might have been a better understanding between them. And with a better understanding, greater ability in Jonathan to cope with the urgings of youth, especially in a place so seductive as Cape Island.

But none of this had happened. None of this closeness had been planned for by Harrah with the thoroughness and detail he reserved for the preparation of a new season. Only Willacassa seemed prepared for the life ahead of her. And here Rachel probably deserved more credit than he did. For the two were always talking, silent Rachel being anything but silent with his daughter.

"Well, what did you find out?" asked Harrah as he pulled out a decanter of brandy from his liquor cabinet.

Henry Sawyer waved him off as he slumped into a chair. "I've had more than enough to drink already. I downed a few at the tavern after work."

Harrah poured himself a drink as he waited for Henry to make his report.

"The girl you asked about lives at the Centre Hotel with her father."

"That much she told me."

"But she's sleeping with a lawyer from Baltimore."

"In the same hotel?"

Sawyer nodded.

"How do you know?"

"The chambermaid told me. She also told me that he frequents Riddle's Tavern. So I had a few drinks there. It wasn't hard to pick him out. He likes to brag about the cases he's won and the girls he sleeps with. I don't know what he likes more, beddin' the girls down or talking about it."

"And you're sure it's the same girl."

"There aren't too many Marcy Crittenfields around, are there?"

"I'm afraid not."

"Her father's a railroad man in Camden. At least that's how he presents himself. But I'm not so sure."

"What do you mean?"

"His clothes are on the shabby side. Shirt collar's frayed and cuffs worn. He may be one of the speculators you see so much of on the Island, looking to find a haven among the rich."

"And Jonathan?"

"He has only eyes for the girl. I saw him up there while I was replacing the floor boards. But he walked right past me. The girl knows how to get him excited. She lets him get close but not too close. Must be tired, I guess." With that Henry Sawyer broke into a good-natured laugh that filled the room.

"What does she see in him, Henry? As young as he is and all."

"Don't know. Unless it's the fact of Trescott House. Jonathan always lays things on a bit thick. To hear him talk, you're making big money here. And he's the son of the proprietor of the place."

"I see. Then it's as simple as that."

"I could be wrong, Mr. Harrah. But that's how it appears to me."

As soon as Henry Sawyer left, Harrah went up to his son's room. It was a tiny chamber with a view of neither the Ocean or Columbia House. The only thing his window looked out on was the icehouse and the stable in back. Harrah realized that in this regard he had treated Willacassa quite handsomely, giving her a larger bedroom with a view of Columbia House and, if one strained, a glimpse of the sea. The best rooms, of course, were reserved for the guests.

But small as it was, Jonathan's room bore the mark of occupancy. Large drawings of ships hung on the walls and a ship in a bottle that he had constructed when he was fourteen sat on Jonathan's writing table. An 1850 Map of Cape Island was nailed to his door. And a spy glass, a ship's sexton, a captain's cap, and a ship's compass cluttered his window seat. A handful of books on navigation lay piled on the floor near his bunkbed which had been removed from a wrecked ship and installed in his room a few years back. A few railroad circulars lay on the blanket of his bed.

Harrah did not know to what extent his son read the books he owned. Nor did he quite understand his sudden interest in railroads. Certainly there was no road near Cape Island as yet.

He removed the circulars and lay down on the bed, waiting for his son to return home. It was near the dinner hour and, if he returned home at all, this was the time Jonathan would put in an appearance.

Studying the maritime sketches on the wall, Harrah was convinced more than ever that the boy in Jonathan had never been squelched. Had he really been taken with the sea, he would have signed on with one of the steamboats that plighed Cape May's waters as soon as he came of age. Instead he was endlessly engaged in such playful pastimes as archery, ten-pins, pistol shooting, and quoits. It was as though summer had a stultifying effect on the boy and no growth could take place until the tents of summer were removed.

As Harrah was about to doze off, Jonathan let himself in.

"Hello, Dad. Didn't expect to find you here."

The boy seemed so happy Harrah almost did not have the heart to spoil things. He got up from the bed and assumed a more authoritative posture.

"Thought I'd have a few words with you."

"About Marcy?"

"Yes, she was here, you know."

"She told me about the visit."

"Well, Jonathan, what's all this about wanting to get married?"

"It's true, Father."

"But you're just a boy."

"I'm eighteen."

"Exactly. Even at twenty-one, you'd be much too young."

"But I love her, Father."

"Love her? You only know the girl a couple of months. Hardly a long-time acquaintanceship."

"Long enough to know I want to be with her."

"Well, it's out of the question," said Harrah, not wanting to but turning his back on the boy.

"Out of the question? Why is it out of the question?"

"For one thing," said Harrah, facing him once more, "you have no source of income."

"Her father will take me into business. All I need is money to invest. Only a small sum. He'll lend me the rest."

"What kind of business?"

"Railroads. That is, he builds equipment for railroads. Everything from carriage cars to steam engines."

"That hasn't been established yet."

"What do you mean?"

"You have only his word for it."

"His word is enough for me."

"Let's forget about her father for a moment," said Harrah. "Let's talk about Marcy."

"What's there to talk about. You've seen her. Tell me she isn't pretty."

"She's pretty enough," Harrah conceded.

"And she wants to marry me. It's not just my idea."

"I'm sure she does."

"Then what's the problem? Why are you being so harsh with me?"

Was he being harsh? Harrah thought not—at least until now.

"I have reason to believe she's known other men," he said.

"I know that." Jonathan uttered these words with a weariness that belied his age.

"And it doesn't bother you?"

"Why should it? I'm the one she has now. I'm the one she comes to for affection. I'm the one that touches her."

"You've become quite the man, haven't you?"

"Well, it's natural to want to touch a pretty woman. And Marcy's as pretty as they come. I'd die if I couldn't lay hands on her."

"Would you die if she were sleeping with another man?"

Jonathan stepped forward and seized his father by his coat.

"Take that back!" he screamed.

"I can't, Jonathan. It's true. She's sleeping with a lawyer at the Centre Hotel.''

Jonathan looked at his father with so much pain in his eyes that for a moment Harrah feared for his sanity.

"Who told you that?" he demanded.

"Never mind. I know. That's all that matters."

"Another man? A lawyer? But why?"

"It makes no difference why. It's enough that she's not exclusively yours—as you were led to believe."

"You mean she's a trollop."

"I prefer my description to yours."

"But that's what she is, isn't she?"

Harrah nodded.

"Well, it makes no difference. I love her anyway. I guess I suspected all along. But it didn't change things."

"It's got to make a difference, Jonathan. If you're the man you say, it can't help but make a difference."

"I don't care two pins about that. So I'm not a man of honor!"

"Forget about honor. It's just not acceptable, Jonathan. Even you must see that, my boy."

Jonathan's eyes clouded up and in time his face fell into the creases of the crying.

"Oh, Father, what am I going to do? I can't bear the thought of not being with her, of not ever touching her!"

"Can you bear the thought of her being touched by someone else?"

"No!"

"Then drop her, Jonathan. Drop her and be the man you say you are!"

Jonathan's condition was so wretched that he could no longer talk. His throat choked on his words. And the tears that streamed down his face all but suffocated him. He lifted up his hand to halt all further talk on the subject. And when Harrah complied by starting to leave the room, Jonathan fell on his bed and dug his head into his pillow.

1854

COMING OF AGE

DEATH

1854 proved to be a dismal year for Harrah. In *The Camden Democrat*, one of the newspapers regularly delivered by steamboat, he read of the passing of Jesse Richards, the Ironmaster of Batsto. After years of failing health, Jesse had finally succumbed. On June 17, at eight o'clock in the morning, the obituary had said. So even giants were mortal. And Jesse had been a giant among men.

Too bad Harrah was not at his funeral to give the eulogy. For there were only two men, outside his father, that Harrah ever really respected. One was the "Mayor" of Burnt Tavern Road, the old smuggler, Bill Hostetler—Cassandra's step-father, may he rest in peace! And the other was Jesse Richards.

For years Richards and iron were synonymous in the Pines, just as the Jersey Pines and iron were synonymous before Pennsylvania iron (and anthracite) did them in. But times and fortunes change, so quickly that one does not always see what is happening. And the Pines fell into troubled times. Somehow Batsto held on, thanks to Richards's perseverence and ingenuity. He imported pig iron to make the stoves and pipes and hollowware that bore Batsto's imprint. But it was a losing battle and in lesser hands Batsto, too, would succumb.

But more than a great ironmaster and a skillful businessman, Harrah remembered Richards as a friend. Hadn't he helped Cassandra through a difficult personal crisis? Hadn't he arranged for Cassandra and Harrah's wedding at the little white church at Pleasant Mills? Hadn't he secured large sums of money and given his personal notes when Cassandra announced her daring plan to build a new irontown which they later named Trescott? And hadn't he been there to cushion the blow when hard times fell upon their venture in the Pines?

Harrah had not seen Richards again since taking leave of the Pines with Jonathan and little Willacassa. At least a dozen years now. It was true that every year Harrah made a pilgrimage to Cassandra's grave at Pleasant Mills. But so sad were these visits that he never had the heart to go the two extra miles to Batsto. Still, for a while, the two men had corresponded, though less frequently as the years wore on. And at Christmas there were always gifts, sent by stagecoach, for the children. Now that Richards was dead, maybe a sentimental journey was in order.

"Let me come, too," said Willacassa. Quite the young lady now, past sixteen, how could he refuse?

"All right," said Harrah. "It's time you visited your mother's grave. She's buried in the same cemetery, you know. The one that holds the Richards's family plot.''

When they arrived at Batsto after a long and tiresome journey, they found the ironmaster's house closed up, Thomas Richards, Jesse's son, having gone to visit his sister in Mount Holly.

"We'll go straight to Pleasant Mills," said Harrah. But he drove his carriage at what seemed a leisurely pace, a growing sense of reluctance holding him back. He had promised Cassandra he would bring the children to see her every summer. But in all that time he had not taken them once. Oh, there were reasons for not making the trip. They were too small to travel. It was a sad journey and he wanted to spare them unnecessary grief. But none of these were the whole reason. At bottom was the plain fact that he wanted to keep them from the Pines. It was as if he were afraid the Pines would seduce them and hold them captive.

As they neared Pleasant Mills, Willacassa suddenly took his arm.

"I sense her presence," she said.

"I'm not surprised. This was your mother's world. This is where she wanted to be. Yes, I sense her presence too."

As the old white frame church with its two entrances came into view, Harrah turned his carriage onto the church road. He had gone about halfway when he stopped the wagon.

"There it is,''he said, pointing to the right, "there where the four cedars stand guard.''

He disembarked and helped Willacassa down.

The cemetery grass was tall and as yet uncut but the stone, lying flat, was plain to see.

"Cassandra Trescott Harrah
1810-1841
Beloved Mother, Beloved Wife
Daughter of the Pines''

"That's a beautiful sentiment, Father," said Willacassa. "I'm glad the word 'daughter' is there. It keeps her young eternally.''

She *was* eternally young, thought Harrah. People like Cassandra never grow old. They fix a time and place in our minds forever. And neither they nor the world that made them ever change.

Willacassa kneeled down and lay the flowers she had brought with her from Batsto on her mother's grave. Harrah stood to a side, allowing Willacassa her moment alone with Cassandra. He watched as she ran trembling fingers over the stone, almost as a grave rubbing.

"Oh, mother, mother!" Willacassa cried suddenly, sobbing out of control. She lay her head on the stone and began kissing the grave. And only when Harrah took hold of her shoulders in a clutch of reassurance did she pull back. A moment later she took her father's arm and was herself once more.

"I never knew her, Daddy. Yet I feel I knew her all my life. And I miss her so."

Harrah now understood why he had been so reluctant to bring Willacassa to see her. Life on Cape Island was a world apart from anything the Pines had ever known. It was an open, boundless, sunfilled world that took its lumps in broad daylight. It knew none of the gloom and loneliness and overgrown wildness of the Pines. Harrah was afraid Willacassa would come in contact with a certain morbidity in the place Cassandra called her home, a contact made doubly morbid by the wrenching away of her mother when she was no more than four years old.

"We were married in that church," said Harrah, looking to draw Willacassa toward happier times. "All of Batsto was there to see it. Jesse Richards saw to that."

"Do you think I will marry one day?"

"What a silly question to ask."

"I sometimes doubt it."

"Of course you will, Willacassa. And you'll be happily married, too."

"Were you happily married?" she asked.

"I was as happy as any man had a right to be."

"Then why didn't you marry Madeleine Culpepper?"

"There are some things it's best not to talk about."

"Especially as we're near Mother's grave?" There was a hint of mischief in Willacassa's suggestion.

"That will do for want of a better reason."

But Harrah was pleased that his daughter spoke of the two women interchangeably. Cassandra Trescott and Madeleine Culpepper. He shook his head in lost wonder. Though he had neither woman now, he considered himself fortunate to have known either one at all.

CHANGE

1854 was also a dismal year for Cape Island. Oh, a few good things were started, like a thoroughfare between Cape May and Cape May CourtHouse. And a road to Town Bank—by a bona fide road construction company, not a gang of farmers working out their taxes. And on the safe side, Congress finally authorized captains of rescue boats at two hundred dollars a year, though the crews were still voluntary.

And there were other interesting developments. The first meeting of the duly-constituted and chartered Masonic Lodge of Cape Island was held in the Evening Star Rooms of Odd Fellows Hall on Franklin Street, with all ten charter members in attendance. And, of course, the main house of Congress Hall was rebuilt, a palatial structure of huge proportions, on the spot where the former building had stood.

But on July 1 the first Camden train rolled across Absecon Beach, some forty miles to the north, into newly-built Atlantic City. What would the new city, with its railroad through the center of town, bring? And how would it affect the economy of Cape Island? It was one thing to be the oldest resort on the Atlantic Coast. It was another to remain the Queen of the Resorts, as Cape Island had been for so many years. The hotel owners were not timid men at heart, afraid of a little competition. But on the subject of Atlantic City they were downright pessimistic.

"I tell you the place will do us no good," predicted Israel Leaming. "The railroad will bring people in from Camden and Philadelphia. We draw large numbers of visitors from Philadelphia. It's got to hurt."

One of the McMakin brothers of the Atlantic House disagreed. "It's a barren beach. I've seen it. The only inhabitants are mosquitoes and green-headed flies. And there can't be much more than a dozen structures there right now."

"There'll be more," insisted Dr. Marcy. "There will be big hotels and amusement houses. And a steamer landing, too. Mark my words."

"But there's no place to visit once you're there. At Cape Island a person can drive to Diamond Beach and Poverty Beach. There's the steamer landing and the old landing at Higbee's Beach. And Cold Springs. And the Lighthouse. And there are few mosquitoes or green-headed flies at the Cape."

"We know that. But do the tourists? That's not what's mentioned in the newspaper advertisements—any more than we take notice of our dusty streets. No, Atlantic City is going to hurt us. We may not end up a ghost town. But the good old days are a thing of the past.''

''Nonsense,'' said Lilburn Harwood. ''So long as we cater to the rich, Cape Island will thrive. Bring the railroads in and people who won't stay long enough to get sand in their shoes—then, and only then, are we in trouble.''

If Lilburn Harwood the hotelkeeper was pessimistic about the railroads, other men in drawing rooms were not.

''I maintain that a railroad from Millville to Cape Island would give Cape May the one thing it lacks,'' said Otis Landborn, a guest at Trescott House.

''What's that?''

''Contact with the rest of the state by land. Right now, except for an occasional stagecoach, it is totally dependent on the sea.''

''It's just as well,'' remarked McMurphree, his tablemate and card partner. ''Have you seen what the rest of South Jersey is like? Marshes, scrub oak and pine, scratch patches of farmland. Take away the shore and you've got nothing but a wasteland.''

''For now. But in time—''

''In time it'll still be a wasteland. I say North Jersey should secede from the South. And Cape Island should be made part of Dixie.''

''How are we going to preserve the Union if we don't preserve New Jersey?''

''The Union won't be preserved. It'll be torn asunder by the slave holders and the abolitionists. Missouri Compromise or no Missouri Compromise, we're going to see Bloodshed.''

''I say the Union will be preserved.''

''Nonsense.''

''I'll tell you what will preserve the Union,'' said Landborn. ''Not the Missouri Compromise. Not the Fugitive Slave Act. But the railroads.''

With a name like Landborn, thought Harrah, how could he not support the railroads?

''A network of railroads across the country will bind this nation together like a cat's cradle. No one part will be able to do without the other. The South will ride cotton and agricultural products to the North. And the North will send manufactured goods to the South. In time, at a much cheaper rate than England or France is charging for an ocean voyage.''

''But the railroads claim they are losing money.''

''On passengers, yes. They'll always lose money on passengers if that's all they run the trains for. You have got to accommodate passengers. Build fancy cars for them. But freight's another matter. All you need is a box car—lots of box cars—and an engine to pull them along. And if you have an accident somewhere along the line, the freight is not a total loss. Some of it spills over, some of it spoils. But the damned cargo is not lost at sea."

"But Landborn, Europe is building railroads. And I don't see that this is preventing border clashes or even open hostility."

"The trouble is Europeans don't speak the same language."

"And we all do!" mocked McMurphree in his best Southern accent.

A roar of laughter shook the drawing room.

A LETTER

1854 was also the year Harrah heard at last from Madeleine.

From the time Madeleine Culpepper left Cape Island in March 1851 to return to her children in Virginia, Harrah had gotten no word from her. Nor did he write. What more was there to say? Hadn't they wrung each other out with pleadings and arguments and declarations of love? Hadn't they made an end of their *affaire*? (He hated the French word for what had taken place between them, but he could think of no other that captured the blissful companionship and joy he had known. So *affaire* it remained, though he hated the word with a passion.)

And then one morning when he picked up his mail, he found her letter among the business communications and the pile of envelopes addressed to his guests. He wanted to tear the letter open at the post office and, with his heart palpitating, read it. But he felt word from Madeleine warranted a kind of privacy the post office could not afford.

It was only when he returned to Trescott House and shut the door of his small office that Harrah finally opened the precious piece of mail and read it.

"Dearest Nathaniel,

I know you will wonder why I had not written until now—just as I have wondered why I have received no letter from you. But the truth is I was afraid to write, afraid of opening my heart to pain again. I will tell you once more, then I will forever be silent on the subject.

I do love you, Nathaniel. But knowing how futile this love has been, I have put considerations of it aside.

The fact is, I married on June 15. I married a man called Benjamin Craddishaw.

Yes, he's from a distinguished Virginia family.

Yes, he will manage my plantation. (He has substantial business interests of his own but will find the time to do this.)

And yes, he will be a father to my children—all the things you would not or could not do.

But I am not writing to reprimand or scold you. For I fully understand what kept us apart. Would you believe that for all the pain it has caused me I respect you for what you did—or chose not to do?

No matter. I am writing to ask you to be my friend. It is doubtful we will ever meet again. And I can bear that—if only I can rely on your friendship. A letter now and then is all I ask. Just a few words. If you will do that, I will be eternally grateful. I will do honor to the name of Craddishaw and perform all my wifely duties. But I cannot bear the thought of never hearing from you again.

May the Lord forgive us for not doing what we should have done, Nathaniel. May the Lord forgive me for putting expediency before love—even if there was an unbridgeable gulf between us.

Madeleine''

Madeleine married. What a devastating piece of news! The world coming to an end would have been easier to take. Harrah reeled before the blow. All his winter dreams, his summer fantasies—shattered by a single letter. He put his head down on his arms, and saw nothing but gloom and waste ahead of him. Summer after summer the same dreary routine. Winters empty of people, with raw and biting winds where hope had been.

Then a moment of rare bitterness. Why the hell did she have to do it? "Whither thou goest, I will go," said the Bible. Why couldn't Madeleine play the part of Ruth instead of an independent Southern bitch with her dependent pups nestling around her? It did not matter that he too was overprotective with his children. She should have lodged where he lodged. Cape Island should have become her island.

"You look like you lost your best friend," observed his son that afternoon.

"What do you want, Jonathan?"

"I need a few dollars, Dad."

"What for?"

"Does it matter? You always say I should never go round without a dollar or two in my pocket."

"Here's two dollars then." Harrah pulled some money out of his billfold.

"I need twenty."

"Twenty! Why, that's more than three weeks' pay for some men."

"I'm nineteen, Dad. I'm not a kid anymore."

"What has that got to do with my giving you twenty dollars?"

"There are girls I want to escort about town. There are things I want to do."

"The Kursaal again?"

"Well, it's the only entertainment around here. During the day, that is."

"What's the matter with the beach? You don't have to spend money on the sand or the ocean."

"That's for mothers or governesses with babies to take care of. Besides, I've been to the beach. My neck is half toasted already. Do you want me to get sunstroke?"

"No, but I think it would be a good idea if you got a job for the summer."

"Doing what?"

"Clerking a store. Digging a road. Any number of things."

"Like hammering nails? I know you spoke to Henry Sawyer about taking me on."

"Well?"

"There's no real money in carpentry. I don't want to waste my youth being an apprentice. I've met some railroad men. They've taken a liking to me. They've offered some interesting propositions."

"What kind of propositions, Jonathan?"

"Of a speculative nature."

"But you have no money to speculate with."

"That's true. That's why I want an allowance."

"You do some work around here and you'll get wages. The only thing I've ever seen you do any work on are the fishing boats going to and from the Landing."

"Boats are all right, Dad. But railroads are infinitely more fascinating. Did you know they've pushed a road through to the Atlantic? All the way from Camden? They've got a small city there now."

"I know all about it."

"One day they'll extend a railroad to Cape Island."

"I'm sure they will."

"Well, I want to be on it. I want to be in the engineering seat when it happens. Meanwhile, can I engineer a loan?"

"Twenty dollars?"

Jonathan nodded.

For some inexplicable reason Harrah thought of Madeleine just now. Would she have approved of Jonathan? Would she have criticized his handling of the boy?

"And when will I get it back?" asked Harrah, peeling off the bills.

"In time, Dad. In time."

JONATHAN

"Do you want to see me, Father?" Jonathan used Father and Dad interchangeably. Dad, when there was an exchange of affection between the two; Father, when their discussions were formal and criticism on Harrah's part was implicit.

"Yes, Jonathan." Harrah tossed a letter upon the flat office desk and watched it slide towards his son. "My Philadelphia bank informs me that I'm overdrawn."

"Overdrawn?" Now nineteen, Jonathan had the self-possession of a young man, but gave only a hint of maturity. Though he had Cassandra's light eyes, Jonathan was a reasonable facsimile of Harrah at the same age: rather tall, quite thin, with broad, even shoulders and a high forehead. His stance was looser than Harrah's and his gait more uncertain. But this was in character with a personality and attitude much less disciplined than his father's.

"It appears that a deposit of fifteen hundred dollars was not made in August. And a second deposit of five hundred dollars not made in September. You carried both deposits to Philadelphia, didn't you?"

"I did."

"Are you sure?"

"Of course I'm sure." Though his eyes were steadfast, Jonathan did shift the position of his feet somewhat.

"You can sit down, Jonathan."

"I'd rather stand."

"All right," said Harrah, abandoning his chair to stand himself. "I would like you to reconsider what you've just said."

"Reconsider?"

"I want you to tell me the truth, Jonathan."

"But I am telling you the truth." Not once did he blink an eye or shift his gaze.

"Then where are the deposit receipts?"

"The deposit receipts? Where I left them."

"And where's that?"

"On your desk. Maybe Harriet moved them when she dusted your room."

"There were no deposit slips, Jonathan."

"What do you mean?"

"You never made the deposits."

"But I did."

"Must I shame you into telling me the truth?"

"Shame me? How?"

Harrah pulled another letter out of his coat pocket. "Did you ever hear of the West Jersey Railroad Carriage Company?"

"Yes."

"Have you invested money with them?"

"Yes."

"A good deal of money?"

"Why do you ask?"

"They want you to invest more."

"Was that letter addressed to me?"

Harrah nodded.

"Then you had no right reading it."

"Under the circumstances I did."

"What circumstances?"

"The fact that you invested my money in the company and not your own. The truth is, you had none of your own to invest."

Jonathan weighed the import of Harrah's words and decided not to comment one way or the other.

"Jonathan, do you admit investing my money?"

"It was my money."

"Where'd you get it?"

"At the Blue Pig."

"Gambling?"

"Yes."

"You're lying, Jonathan."

"I'm not lying."

"Oh, I'm sure you've gambled at the Blue Pig. But you didn't win, you lost. And the proprietor said you weren't welcome there anymore."

"There are other places to gamble."

"Still, it was my money you invested in the Railroad Carriage Company, wasn't it?" Harrah placed his hands on Jonathan's shoulders and shook him as he said this. But his son did not respond.

"Tell me Jonathan. I want the truth. No more lies!"

Still no response.

"Did you know," said Harrah, releasing his grip on him, "that the West Jersey Railroad Carriage Company is a fraud? That it no more has railroad carriages than you or I?"

Jonathan's eyes gaped unrestrained.

"Where'd you hear that?"

"It's in all the newspapers. This letter must have been a last-ditch effort to raise some money. The officers of the company have since fled."

Jonathan now took the chair Harrah had offered him earlier.

"They can't do that," cried Jonathan. "All my money is tied up in the company!"

"My money, you mean."

The young man nodded, then broke into a fit of sobbing. "I was going to pay you back, Father. But the return on the money was so big at first, I just kept pumping money into it."

"Why didn't you come to me? I would have checked them out," said Harrah, not a little moved by his son's anguish.

"I wanted you to be proud of me. I always thought you were too cautious in business, too conservative."

"For good reason, Jonathan. For good reason."

"What are you going to do with me?" Jonathan looked at his father through a curtain of tears.

"I don't know yet," said Harrah.

"But it will be severe, won't it? The punishment will be severe."

"You've squandered a good deal of money. What do you expect me to do? Let you off scot-free?"

"Oh, do what you want. Arrest me. Horsewhip me! I don't care!" cried Jonathan.

"Horsewhipping you won't bring my money back." With this comment, Harrah left his son sitting in the office.

The next morning Harrah was full of resolution. He would speak to Captain Whilldin. The Whilldins had been at the Cape for as long as anyone could remember. Wilmon Whilldin, Sr., had been a renowned steamboat captain until his death in '52, and Wilmon, Jr., was carrying on the tradition. The captain knew Jonathan and was fond of the lad, and the feeling was reciprocated. In his hands Jonathan would be fashioned into a responsible seaman, and then one day the young man would captain his own steamboat. Though he would make port regularly, discipline was what Jonathan needed, and aboard ship he would have to tow the line.

Harrah could not wait to tell Jonathan of his plan. But only a dismayed Willacassa appeared at the breakfast table.

"He's gone, Father," she said.

"Jonathan?"

"Yes, he left a note. Did you quarrel?"

Harrah tore the note open. But even before he read it he was berating himself for having waited. He should have settled the matter while he still had Jonathan in his office. He should have shown that though he disapproved of what Jonathan had done, he still loved him.

"I know you won't forgive me," the note began in a hurried scrawl. "And I'm not asking your forgiveness because one day I'll pay you back. I'm sorry things had to turn out this way. I wanted you to be proud of me. But it's hard being your son because you're always so righteous—though you're not a bad father, really. Don't look for me. I'm nineteen and old enough to take care of myself. Tell Willacassa I'll miss her. Rachel, too.

Love,
Jonathan''

1855

NEW ACQUAINTANCE

EXPECTATION

Willacassa greeted the New Year with determination if not resolution. She was determined that 1855 be a happy year, a marked change from the year before. Having turned seventeen in December, she looked toward summer for the fulfillment of her expectation. Her expectation, of course, was to meet a handsome young visitor who met all the requirements of a serious year-round suitor.

But though she had her share of beaus and attended many a hop when she could get away from her duties at Trescott House, she met no one who approached the standard she had set. There was an on-going conflict within Willacassa. The unadorned realist in her, more a leftover from her childhood and childhood's honest observation, was doing battle with her desire to break out of the mold of her experience. For experience had taught her that snappy dress could hide shabby dissipation. And the clink of mint-julep glasses and the fragrance of imported seegars could give way to drunken outrage and reeking pursed lips where no invitation to kiss had been given.

And so she did all the things expected of a pretty young lady native to the place. She took promenades along the salt-sprayed strand. She escorted her present host to the Cape May lighthouse with its revolving light. She visited Cold Spring and sampled its mineral water. She even pointed out the dwelling of the albinoes who years before had been on display at a Philadelphia exhibition.

Willacassa recognized that in some ways things might be difficult for her. Having spent most of her childhood on Cape Island, she knew the ways of the summer visitors, especially the young men and women. She recognized both their charms and their foibles. As an unadulterated critic, with no self-deception in her makeup, she found it hard to play the game of summer make-believe.

Still, she made the attempt and even began to see Cape Island, not from the inside out but from the outside in. As a newcomer might, arriving for the first time from the Landing and encountering the red-top houses and the widow's walks and the church steeples, she took note of the flying flags and the archery tents and the ten-pin alleys and the dolled-up conveyances that rumbled about the town. And always in view down the side streets she saw the ocean, its waves lapping the beaches and its breakers delighting the swimmers as they emerged from the portable bath-houses to test the waters.

But when summer drew to a close and the wagons and carriages loaded with "guests" jammed the Turnpike on their way to the steamboat landing, Willacassa was no nearer fulfilling her expectation than she had been prior to the ten-week season.

"I hate the passing of the season," she complained to her father as he sat reading the Philadelphia newspapers and *The Ocean Wave*, a local publication which Colonel Johnson was putting out on a trial basis.

"I love this place. And it seems the center of the world in summer. But off-season I have the desire to be elsewhere, anywhere. I can't bear the emptiness—almost all the houses shut down and most of the stores boarded up. Even Jonathan hated it. And that's one reason he ran off. Am I an ingrate, Father? I know this place is paradise, yet I want out of it!"

"An ingrate? No. You're young and can't bear the quiet. Quiet is the bane of youthful existence. But when you reach my age, you'll welcome the change. Of course, you've got a way to go!"

"I'll go stark raving mad if something doesn't happen soon. Anything—a hurricane, a shipwreck, even murder!"

"I'll see what I can do," smiled Harrah, putting his newspaper down. "Maybe President Pierce will honor us with a return visit. I'll send him a letter by morning post."

"Something like that," laughed Willacassa in spite of herself.

Then abruptly all of Willacassa's complaints ceased. Not only did she refrain from her criticisms of Cape Island, she began suddenly to hum little ditties that she had heard at the Kursaal during the summer and without knowing what it was sang a popular aria from *Don Pasquale*, inventing her own lyrics for the melody.

It was Rachel who inadvertently spilled the beans.

"Where's Willacassa?" Harrah had asked, having missed the girl at the dinner table.

"Walking past the Mansion House, I'll venture."

"The Mansion House? But it's closed for the winter."

"There's a young man looking after the place. He lives in a front room. A sort of caretaker, he is."

"A caretaker?"

"Least that's what I've heard."

"Well, what does Willacassa have to do with that?"

"Wants him to notice her, I expect."

"Notice her?"

"Just as she's noticed him. He's a handsome young man. But a mite too shy and quiet for Willacassa's tastes."

"How do you know all that?"

Rachel did not crack a smile. "She told me so herself."

Harrah rose from his chair and paced up and down the drawing room.

"Then why is she going out of her way for him?"

"Who else is there for her to go out of her way for?"

Who indeed? Cape Island wasn't noted for its young swains in winter. Most eligible young men were gone by September, back to Maryland or Delaware or Virginia or wherever else they came from.

Besides, Harrah wasn't sure he wanted Willacassa to be looking for young men just yet. She had known her share of summer beaux. Ten weeks of frolicking—when she could get time off from Trescott House—was more than enough for any girl. It took all winter and then some to get foolish summer notions out of one's head.

But Harrah did not share his feelings with his daughter. And Willacassa continued her vigil. Not once, however, did the new man in town emerge from the Mansion House at an opportune time. And when she did meet him it was quite by accident.

One afternoon as she was returning from Schellenger's Landing on an errand, she saw him on Washington Street with a tripod and a transit.

"Are you the new surveyor in town?" she asked.

"When I get the work. Thomas Quigg is planning to buy this house, then move it. He wants to be sure it will still stand on his property."

"A sensible notion." She said no more, for she noticed for the first time how truly handsome he was, with his well-chiselled features and dark hair and lashes.

"Here," he said, not aware of the interest he stirred, "you can be my rod man for a moment." He handed her a tall staff and directed her some fifty feet back where a Spanish oak with a blaze on it stood. "Now hold still," he said. "I've already done this once, but I'm rechecking."

Willacassa did her level best.

"Thank you," said the surveyor when he finished making notes. "You made my job a lot easier. You don't know of any other property that needs boundaries, do you?"

Willacassa thought a moment. "Some lots on Congress Street perhaps. Everyone wants to build cottages like the Steiners."

"Who owns the lots?"

"My father might know. But the Court House has all the records."

"Thank you, Miss—"

"Harrah. Willacassa Harrah."

"Thank you, Miss Harrah."

143

"And your name?"

"Eliot."

"Eliot what?"

"Just Eliot, that's all."

Willacassa thought his insistence on a single name a bit unusual, if not impolite, but she did not press the issue.

"I understand Mrs. Corgie makes a fine breakfast," said the surveyor, slinging his equipment over his shoulder. "Would you care to join me? You've earned it in a way."

Willacassa had already had breakfast, but she agreed without hesitation.

"What brings you to Cape Island at this time?" asked Willacassa when they were seated at the smallest of Mrs. Corgie's three tables. "Off season is so desolate here."

"Mr. Ludlam decided to have a live-in caretaker. A caretaker lowers the insurance rates. I had written for lodging not knowing he was closed for the season. He made the offer and I accepted."

This did not answer her question. But Willacassa decided to forego pursuing it.

"Besides," he continued, "it's not as desolate as you say. I understand Captain Kidd's treasure is buried near here."

"At the Landing or Higbee's Beach," smiled Willacassa in spite of herself. "But I wouldn't put much stock in it."

"And there appears to be building going on all the time. Even in winter, I hear. Good for my trade. Though it's old houses that really interest me."

"Old houses?"

"Yes. Once the season's over, I'll need a place to live. I can't afford your summer rates."

"Few houses are really that old. And those that are available—especially outside of Cape Island—are virtually abandoned."

"Well, no need to trouble about it now. Summer is a long way off."

"Meanwhile you've got a place to live."

"Yes. And I can tell people back home that I stayed at the Mansion House."

"There's something to be said for that." She did not dare ask where home was.

"But I hate playing the caretaker," he added.

"Too much to do?"

"Too little. I'm tied down to the place. I don't have the freedom to do what begs doing. I made no agreement to be on the premises at all times. But that's pretty much what's expected of me."

"Is freedom so important to you?"

He took a long look at Willacassa through his dark lashes and nodded "Without freedom, life is meaningless. Mere existence, no more. Doesn't matter who you are."

He was so intense when he said this that Willacassa remembered almost nothing else from their meeting.

SOMETHING OF A CONTRADICTION

Eliot was something of a contradiction for Willacassa. For several weeks his comings and goings followed an almost predictable pattern. He was seen walking on the beach's edge in the morning, taking breakfast at Mrs. Corgie's, and buying newspapers or magazines at Canning's on Jackson Street.

There were even reports that he was seen at Riddle's Tavern at night, but these she tended to doubt. Generally a few roustabouts and ne'er-do-wells hung about the place. Willacassa could not imagine anyone as fine-tuned as Eliot filling his hours there. Besides, whenever she walked by the Mansion House at night, she saw a light flickering in his room, even his shadow moving across the window. Perhaps a beer or two at Riddle's to assuage a thirst. No more than that. For taverns were where small talk and little men prevailed.

But then Eliot would disappear for days on end—vanish totally from the streets of Cape Island. And no one had the slightest idea where he had gone or when he would come back—except that the storekeepers expected him back, for he would order books and newspapers and a smoking tobacco not found in these parts.

Didn't he know there were supposed to be no secrets on Cape Island? Everybody's life was an open porthole. That was why Willacassa herself liked to keep people guessing, to tilt them off balance with sometimes erratic behavior. Yet Eliot was either the most shy or secretive person she had ever encountered. And for a moment Willacassa knew how the rest of the Cape Islanders felt. She too had a squatter's right to know something about the town's new inhabitant, and she wanted to exercise this right—at least as far as good taste was concerned.

She would start with his "disappearing act"—for that's what Willacassa called it. Was it an aberration in an otherwise ordinary existence?

"Where would a young man visit in Cape May County?"

"How do you know he's in Cape May County?" countered Harrah who was pleased his daughter had taken him into her confidence.

"He never goes to the steamboat landing."

"Never?"

"I've checked with the steamboat captains. They've never had a passenger of his description."

"Then there are only two possibilities," teased Harrah. "He's part of a smuggling ring in the Pines or he's got a ladyfriend."

146

"A ladyfriend?"

"Yes, a woman who is unable to see him at the Mansion House."

"Oh, Daddy!"

"Either way," chuckled Harrah, "he'll have a lot of explaining to do."

"Explaining? He doesn't have to answer to me."

"Then there's no point worrying about it, is there?"

CARETAKER

It was Henry Sawyer who broke the news.

"Is your father about!"

"No," replied Willacassa, "he went to Philadelphia."

"Well, there's been a fire at the Mansion House. The caretaker's room. I've got to get some boards and patch it up."

"Anyone hurt?" She tried to be matter-of-fact about her interest in the affair.

"The caretaker. Some burns on his arm. Refused to see a doctor, so I suppose he's all right."

Willacassa gathered up her cloak, "I'll stop by and see if I can be of any help."

"I'll stop by myself as soon as I get hold of some lumber. But I'd hate to be there when 'Uncle Smith' gets wind of this."

"Mr. Ludlam will understand, I'm sure."

"Understand what? That his caretaker fell asleep and let his pipe set fire to the bedding?"

"Is that how it happened?"

"That's how it happened."

Sawyer departed as suddenly as he had arrived.

Eliot's arm was in bandages when he admitted Willacassa to the vacant sitting room of the Mansion House.

"Are you all right?" she asked.

"Things could have been worse. I could have burned the whole place down. Fortunately, only my room was gutted."

"Does Mr. Ludlam know yet?"

" 'Uncle Smith' knows, all right. I'll have to make good for whatever the insurance doesn't cover. But on the whole he took it rather well. I guess he's learned to deal with misfortune over the years."

Willacassa sat down. "You can stay at Trescott House if you're not welcome here any more."

Eliot sensed that this was not an ordinary concession on Willacassa's part and did not know what to make of it.

"That's extremely kind of you. But—"

"You needn't decide at once. Only if Uncle Smith turns you out."

"Aren't you afraid I'll set fire to your place too?"

"Not really. We're closer to the ocean. And the water supply there is more than adequate."

Seeing his bandages unravel, she rose to rewrap and retie them—all of which she did quite neatly.

Eliot smiled—one of his rare smiles, for he was not a smiling man.

"You're quite accomplished for one so young, Miss Harrah."

"I'll be eighteen in December."

"Eighteen? Is that so? And I suppose you'll be marrying soon?"

"Not for a while. At least not until I've seen something of the world. Cape Island is so provincial, don't you think?"

"Every city and town in America is provincial. But few are so charming as Cape Island."

It was Willacassa's turn to smile.

"You're perceived as a man of mystery, you know."

"I am? By whom?"

"By the townspeople. No one seems to know when you'll suddenly disappear, then reappear again."

"Does that trouble you too?"

"A little."

"Well, don't let it bother you. My life is quite mundane, I assure you. I disappear every now and then to give the appearance of things happening. But nothing of importance really occurs."

"I find that hard to believe, Eliot."

She pronounced his name in the hope that he would enlighten her as to the rest of it. But though he noticed and took pleasure in her using his name, he volunteered nothing new.

Looking at the calendar clock on the wall, she prepared to leave.

Suddenly Eliot looked at her without his usual reserve. "I do appreciate you're coming by. It was a sweet, kind thing to do even though I'm not terribly injured."

"It was nothing," she whispered.

He rose and stood before her, his handsome face a mass of concerns.

"I can't say anything now. I've made a mistake and must attend to it. But maybe some day you'll think kindly of me—kindly enought to—"

"Be your friend?"

He nodded, still under the spell of the moment.

"I am your friend now, Eliot."

She closed the door softly and returned to Washington Street in time to see Henry Sawyer drawing up his wagon with a load of lumber.

PANIC

The next time they met was outside the small brick bank on Decatur and Lafayette Streets.

"What's happening?" asked Eliot. "The natives are in a panic."

There was a good deal of scurrying about by hotel owners and tradesmen in the dusty streets and on the sidewalk of the Bank of Cape May County.

"The bank's gone under," said Willacassa.

"Oh, my Lord!"

"You had money there?"

"God, yes! But it's not my money! What am I going to do? It'll take years to pay it back!"

Willacassa had never seen such anguish on a person's face. She had no inkling as to the cause of it. But there was no denying the young man standing before her was a portrait in torment. She wanted somehow to ease Eliot's pain but knew intuitively that there was little she could do.

"It's a state bank, so eventually they'll make good," she said. "But it's a question when. My father heard that in a couple of months all circulating notes can be presented to the State Treasurer for payment. But he suspects it'll be a year or two, if not longer."

"A year or two? I can't wait that long! I'll start for Trenton at once."

By morning she learned that Eliot had packed his things and boarded the steamboat *Delaware*, bound for Philadelphia.

Willacassa was shattered. She could not understand his leaving so suddenly, and without a word of goodbye. It did not seem like Eliot, even though he was unpredictable. And it flattered her not at all as she stood on the Landing at Delaware Bay to have meant so little to him. A girl liked to think men noticed her, particularly men she showed an interest in. She had a way to go if she was going to make her woman's mark in the world.

Willacassa watched the waves wash the pebbly beach, and then looked out at the rough waters of Delaware Bay. But she saw no sign of the steamboat Eliot had taken. She shook her head in abject disappointment. Eliot? Was that his last name or his first name? Now she would never know.

1856

FRANCIS GRANDEE

THE GRANDEES

The Grandees arrived at Cape Island en masse. John Grandee and his Creole mistress took lodging in Trescott House. The rest were ensconced in the magnificent Mount Vernon Hotel which, after much fanfare and several years in the building, had opened the previous season, its mile and half of verandas and balconies wrapping itself around the mammoth structure.

The Grandees were Huguenots and had settled in South Carolina when the colony was still a religious haven. They had purchased land north of Savannah, not too far from Port Royal Sound, and apparently had prospered, operating several large plantations under the family name. At least that was how John Grandee told it.

"Finding the summers too hot in South Carolina, we came by schooner—our own—to Cape May. We feel right at home here. The cooking isn't Southern. But the clientele, at least a good part of it, is."

There was a constant exchange of visiting. Hazon Grandee and his wife, Caroline, would come with their son, Francis, to lunch at Trescott House. And John, when his mistress slept late, would take coffee with them or Hazon's brother, Philip, on the veranda of the Mount Vernon Hotel, the whole facade being as imposing and luxurious as a Venetian palace of the Doges.

Francis Grandee was one of those young men, too pretty (with his dark wavy hair and green eyes) to be masculine, but was masculine enough; too proud to mix with his inferiors, but was affable enough; and too stupid to know that insolence in a young woman was a form of flirtation.

"I've come to see my uncle," he announced.

At the desk Willacassa recognized him but pretended ignorance.

"Your uncle?"

"John Grandee, Esquire."

"Oh, that gentleman, He's not in."

"Not in?"

"No, but his mistress is."

"His mistress?"

"Yes, you know, the pretty Creole girl."

Francis Grandee was too flustered to make an intelligent response.

"Do you want to see her?"

"Well—"

"I wouldn't if I were you."

"Why not?"

"She's too old for you."

"Too old? She's only nineteen. Besides—"

"Too old in the ways of the world, I mean. And too young for your uncle, I might add."

"I don't really think this concerns you, Miss—"

"Harrah. But Willacassa will do. I can see you can't take joshing. I didn't mean to be so hard on you. But you come on so highborn, so sure of yourself, I thought I might take you down a peg."

Her pretty smile eased his discomfort, but young Grandee still did not know what to make of her.

"If you really want to see the girl, you may," she continued. "I'll let her know you're coming so she can put something on."

"That won't be necssary. That is, I really came to see my uncle. I must have missed him somehow."

He started to leave, and then turned round as though in afterthought. "Did you say you were the proprietor's daughter?"

"No. But I am."

"You're not too young to attend the 'hop' the hotel is having tonight, are you?"

"Too young? I'm almost nineteen." Willacassa enjoyed bedevilling him. "Are you asking me to go?"

He nodded.

"Then I'll be ready."

The "hop" was very much a tradition on Cape Island. In the old days guests at their dinner tables would pass a sheet of paper around to see if there was any interest in a dance. If a sufficient number of devotees signed up and expenses could be defrayed, a "hop" was held. Sometimes the affair was advertised and all the other houses invited. Later, the houses on the Island took turns holding "hops." It was a good way to show the places off. By season's end all the hotels had their chance, and their guests got to know who else was staying in town.

But with the increase in the number of hotels on Cape Island and with the advent of such mammoth hostelries as the United States Hotel and the new Congress Hall and the Mount Vernon Hotel, the custom had been chipped away. Now Beck's Band, in all its splendor from Philadelphia, came down around July 15th and remained at the Mount Vernon for the rest of the season. And Saturday night was generally "hop" night for the hotel's guests and visitors.

"Who's coming for you?" asked Harrah that evening.

"Francis Grandee," said Willacassa in her finest white gown, with a flower in her bosom.

"John Grandee's nephew?"

"You don't approve?"

"I don't approve or disapprove. But I would like to know about these things before they happen."

"Well, it just happened and there was no time."

Harrah recognized the refractoriness of his daughter and decided this was not the time to dispute her.

Within the hour Francis Grandee arrived and with impeccable Southern manners introduced himself to Harrah.

"Francis Grandee, Mr. Harrah. My Uncle John is a guest in your house. I have the pleasure of taking your daughter to a dance tonight."

He was not as tall as Harrah would have liked and his vowel accents were a bit too prolonged to suit the proprietor of Trescott House. But altogether the young man, handsomely tailored in a maroon evening coat and white trousers, made a favorable impression.

"You're staying at the Mount Vernon, I understand," said Harrah.

"We all are—except for Uncle John. A magnificent Hotel, I must say. With every service. We have a bath in each bedroom, with hot and cold water taps. Artesian well water, I understand. We eat salon à manger. That is to say, nothing is carved at the dinner table. The wines are absolutely the finest. And the music is first-rate. Even our laundry is provided for."

Harrah acknowledged what he said with a nod.

"Of course, your place is first-rate too. Uncle John said he wouldn't trade it for ten Mount Vernons."

"Even if we have no horse railroad?" interjected Willacassa.

"Horse railroad?"

"To carry passengers from the hotel to the beach. All without charge, I might add."

"Then you know about that?" said Grandee.

"You're not the first young man from the Mount Vernon to ask me to a hop."

"I'm not?"

"No. And I've been to the pistol galleries and archery tents, too. It's very hard to live on Cape Island and not know what's going on."

"Then you will show me those things tomorrow," said Grandee with a bow. "For I haven't seen them."

"Tomorrow? We'll talk about tomorrow only after the hop," replied a mischievous Willacassa.

THE "HOP"

The ballroom of the Mount Vernon Hotel was one of dazzling splendor. Several hundred feet long and at least sixty feet wide, it was lighted by forty ornate chandeliers suspended from the ceiling like the crown jewels of Russia which had been reproduced and widely seen in all their glory in a recent edition of *Harper's* illustrated *Weekly*. No candles here or disagreeable oil. But gas burners that threw clean, bewildering light into as cheerful and luxurious a room as Willacassa had ever seen. And the brightness was made even brighter by the white gowns and bejewelled white necks and arms of the powdered ladies and the white jackets and shining brass instruments and music stands of the members of Beck's celebrated Philadelphia Band. Even the dark complexions of the colored waiters who moved about with trays of sparkling beverages in shiny crystal stemware shone white in the milky incandescence of the evening.

Francis Grandee and Willacassa arrived none too early. As soon as they entered the huge ballroom, which served as the hotel's dining hall during the day, they heard the rhythmic strains of the schottische. And Francis with a short bow engaged Willacassa in dance.

The schottische was followed in rapid order by a waltz, a german, which was more caprice than form, intermingled with passing notes and waltzes, and a quadrille in various rhythms but principally in two-quarter time. A brief intermission, during which a few of the band members were seen quaffing beer in a small room off the dining hall, gave the participants a breather.

Francis Grandee proved to be an excellent dancer. He was not afraid to raise his foot off the floor, nor was he timid about seeking out his partner when exchanges were taking place. He seemed to know the steps involved and what he did not know he carried off with reasonable approximation. Though his movements were not as fluid as one native to the dance, he was never awkward or heavy on his feet.

During the intermission Francis introduced Willacassa to his mother, a cultivated, handsome woman whose hair was prematurely gray, and his father, a rather unpredictable man who was by turns charming and surly, charming as he kissed Willacassa's hand and surly to the waiter who had brought him scotch and water instead of bourbon. Francis' uncle and his Creole mistress Willacassa already knew because of the room they shared at Trescott House.

Before she could engage Mrs. Grandee in conversation or sample the scoops of ices, melons, and pineapples that graced the table, the band once again struck up a waltz and Francis wheeled her onto the dance floor. Hard on the heels of the waltz, the new and popular "Cape May Polka" was played and, as soon as a couple could be found to assume leadership, a cotillion was formed.

Willacassa knew that some people, particularly of the Methodist Church, did not approve of dancing. They regarded it as provocative and sinful, a notion she did not share. To Willacassa, dancing was an expression of high spirit, of innocent gaiety, of freedom. As a young woman she felt the need for a release of emotion and physical energy.

But as Francis Grandee practiced the act she began to have her doubts. During the cotillion, when they changed partners, he was forever looking her way as though he could not wait until he enjoyed his favors once more and had his arm about her waist again. And when they were waltzing, whirling around the dance floor and advancing in wider and wider circles round the room so that the ceiling seemed to be spinning and the chandeliers were in danger of flying off like flaming pinwheels, she felt that he was making love to her.

She did not know how she would have reacted to this in private, but in the public view of the Mount Vernon Hotel's ballroom she experienced an ambiguity of desire and distaste.

"Let's stop a moment," she called, afraid that he might not hear her.

"Stop? But why?" Her partner was scarcely out of breath.

"I'm getting dizzy!"

Francis Grandee gracefully guided her to a side of the ballroom where he found an empty chair.

"Here, try this."

She sank into the chair and began fanning herself.

"Would you like something to drink?"

She nodded.

"I'll be right back."

While he was gone, Willacassa took the opportunity to watch the other young women on the dance floor. Not yet nineteen, she dismissed anyone older than herself as not being typical.

For the most part the young ladies that she watched were elegantly attired and coiffed. Though they wore the two dominant modes of hairstyle—one parted in the middle and swept down under and back; the other, Southern fashion, with pinned up curls—they exhibited but one type of gown. And this was off the shoulder and cut low, low enough for bosoms to peep through, at least where there was bosom enough to make a showing.

And their gentlemen partners exploited these favors to the

fullest. They glimpsed and they peeped. They leered and they ogled. And when they were subtle about what they looked for, they wore frozen smiles on their faces and frozen sweat on their brows, fixing their eyes in a frozen stare on the eyes of their partners though they would rather they were fixed elsewhere.

"Feeling better?"

Francis Grandee handed her a glass of punch.

"Yes, thank you." She sipped the punch and put the glass aside. "I'm afraid you're too much for me. I'm exhausted. I'd like you to take me home."

Young Grandee was heavily disappointed.

"But we'll try again sometime, if you want to."

"Certainly."

Harrah was sitting on the Trescott House porch when the couple arrived.

"Everything go all right?"

"Everything was fine," said Willacassa as she bade her escort goodnight.

But Harrah could see that something was bothering his daughter.

A NEW REALITY

What was bothering Willacassa was a new reality that had crept into her psyche. With the advent of puberty she had developed a sexual awareness common to adolescent girls. And she had, she thought, put this into perspective as she grew older. She understood the laws of nature, and she had in fact categorized the men and women she knew. Some she thought of as sexless; most as normal; a few as passionate; and the rest as vulgar.

By far she liked the sexless ones best, both as friends and as acquaintances. Rachel, who had been with the family from her mother's time on, she adored. With Rachel one thought only of shiny, scrubbed kitchen utensils, platters of berry or fruit pie, and a warm bosom to lay her head against when she was ill or unhappy.

And then there was Joseph Leach, Publisher of the *Ocean Wave*, Cape May County's first newspaper, and a good friend of her father's. In appearance Mr. Leach was a spare, long-nosed New Englander with piercing gray eyes and a long beard parted in the middle. Though his marriage and his many children gave evidence of something more than sexual indifference, this delightful man seemed to rise above the flesh and its bedevilment as a daring, brilliant spokesman for the community. When he wasn't scolding Cape Island for its dusty, unpaved streets, he stood as a staunch Republican in fervent opposition to slavery and the spread of slavery. But he was at his best with the little things, complaining of a penny auction store whose crowds made it impossible to pass on the sidewalk or noting that coal and woodpiles had disappeared, that Mr. Green's tavern had been robbed of rum during the off-season, and that a quantity of preserves and pickles had been removed from George Ludlow's confectionery store.

Only Henry Sawyer, who was now married and a father, could be a friend and tolerated for his passionate nature at the same time. Whether he was replacing boards at the Centre House or racing headlong down the beach with one of his "cavalry" horses or simply taking beer with some comrades at Riddle's Tavern, he was always vital and generous and full of the Lord's bounty. Other people might be as slow as coal tar in winter when it came to taking a stand. But Henry was ready to fight fire with fire.

Francis Grandee was vital enough; and with his short, well-shaped nose and molded face, he was prettier than Henry. But Willacassa was afraid his attentions to her were not born of selflessness. Like most comely girls, Willacassa wanted to be admired. And she knew she had his admiration. But young Grandee's readiness to take her round the waist, to whirl her about on the dance floor for all to see, to introduce her to such as his Uncle John, threatened her female equilibrium. And the way he looked at her was schooled not so much out of regard as out of desire, even lust.

What Willacassa feared was exploitation. Someone like Henry Sawyer would no more take advantage of her now than when, as a little girl, she had a crush on the young man. Francis, she believed, would not be so noble if he had the faintest intimation that she liked him half as much as she did.

And so when Grandee came the next day with flowers in his hand and a note wishing that she were better, particularly as he wanted to take her to another hop after the sabbath—it did not matter which day—Willacassa pretended that she was still indisposed and sent Rachel downstairs with her regrets.

Francis Grandee left, but not before telling Rachel that he would be back the next day and the day after that until Willacassa recovered.

"Your young friend was here this morning," said Harrah when she came to the dinner table.

"I know."

"Didn't you want to see him?"

"Not so soon."

"Afraid of disturbing the sabbath?"

"No, not that."

"Why not then?"

"I don't want him to think I am more interested in him than I am."

"Very wise," said Harrah. And in fact he thought it was wise. But he did not for one moment believe his daughter's disclaimer.

THE BEACH PARTY

"How would you like to attend a beach party?"

"Where?" asked Willacassa.

"At the east end of the beach," said Francis Grandee, pointing in the general direction, past the breakwaters, where soaring curlews and gulls seemingly marked the spot.

"When?"

"This afternoon when the tide recedes. My uncle John has a couple of hundred clams ready to be delivered, Cape May pot pie, and all the rest. He's loading the wagon now."

"All right," said Willacassa, "if I can get away. My father wanted me to do some lettering on a signboard."

"The signboard can wait."

Willacassa met the wagons at three. Some eight couples, including Francis' mother and father were waiting to be hauled, like baskets of corn, to the designated place.

"Here, you can ride with me," said Francis, offering to share his horse with her.

"I'd rather go in a wagon," said Willacassa, taking the proffered arm of huge Uncle John and climbing aboard.

The ride by wagon was not nearly as smooth as one by carriage on the hard strand. It seemed to Willacassa as though the wagon's rear wheels found every stone on the beach to ride up and over on, coming down with a heavy thump and jerking forward again until it found the next stone. But as each encounter generated hearty squeals and laughter, Willacassa overlooked the jarring shocks to her body. Seeing her discomfort, Francis Grandee made an "I-told-you-so" gesture and offered to share his horse once more. But each time, with less conviction than before, Willacassa rejected his offer.

When at last they reached their destination, the celebrants jumped down from the wagons onto the sand. As soon as the parties were assembled, Uncle John announced in a booming voice the arrangements for the day.

"First, my brother Hazon, Mr. Potter, and myself will take charge of the clambake. It will be an Indian clambake, something Mr. Potter has experienced firsthand. While waiting for us to set things up, the rest of you will take your swim. The beach to the east is assigned to the women. The beach to the west to the men. My nephew Francis will see that the sexes are kept at respectable distance. And Miss Harrah will help him."

All laughed.

"When I ring this bell, all the ladies will come out of the water to help with the fixins. With Mr. Potter's help they'll set things up on the tailboards of our wagons. A little later the men will join us to share in the eatin'." He paused and surveyed his circle of listeners. "Agreed?"

"Agreed!" they chorused.

While the rest of the group broke up to prepare for a dip in the surf, the clambake committee set to work. On a flat, level spot, where the sand was hard, they opened dozens of clams and placed them shell-hinge up in a circle about six feet in diameter. They covered this bed of clams with small and large sticks and a layer of fine brush that Mr. Potter had brought in from a small cedar grove off the beach.

Uncle John was now ready to fire the clam bed, but his hand was stayed by Mr. Potter.

"They'll be wanting to watch," he said. "So let's wait."

But the last thing on the bathers' minds at this point was the clams. With a rolling surf in front of them, displaying just enough treachery to assure them of a spill or two, they removed their outerwear. And the women in long skirts and short sleeves and the men in their long striped suits waded into the water.

With a strong undertow from the groundswell of the ocean, the ladies insisted that the men stand close by for support. But with all this protection a few of them were still knocked over by the surging waves, and it was decided the bathers would form a ring by holding hands.

Flanked by Francis on one side and his father on the other, Willacassa felt more surrounded than protected. Breaking away, she dove headlong into a rolling wave, coming to surface a full twenty feet behind it, her hair hanging dank and parted fitfully.

Francis, though not a good swimmer, went after her. He regarded Willacassa's defection from the group as impetuous and foolhardy. What had gotten into her? But when she swam beyond the breakers, he turned back to watch.

Feeling she had asserted her independence, Willacassa rode an incoming wave back towards shore. She could see Francis waiting at the water's edge, and this pleased her. But she did not see a huge breaker rush in on her. Tumbling forward, she disappeared for a time. As Grandee scanned the water, which lapped wildly back and forth, Willacassa's form appeared once more and she found her footing. This time Francis caught her hand and pulled her, protesting, onto the beach.

"You had no call to do that!" she exclaimed. "I need no one to help me!"

"I was anxious about you, Willacassa. I don't know why you're snapping at me!"

Willacassa coughed water, but before she could reply, she heard Uncle John's bell ring and his booming voice call everyone to dinner.

It was only when they had changed back into dry clothes and stood with the others in an expectant circle that Willacassa was herself once more. But she still made no concession to Francis. Instead she watched the clambake committee set fire to the fine brush covering the clam bed. Getting down on his huge haunches, Uncle John set a match at one end, and Mr. Potter set his on the windward side at the other end. Soon the bed was crackling with fire, the dry weather of the past few days enabling the brush to go up like paper in a fury of finger-writhing, sky-soaring flames. To Francis Grandee the small fire had all the roar and smoke of a forest burning.

Francis Grandee turned to Willacassa for some sign of emotional sharing in the spectacle. He wanted some kind of romantic recognition from her. But she had eyes only for the fire, her face tinted red by the hot glow.

Mr. Potter, whom Willacassa now recognized as a local fisherman, and his wife set up a triangle of poles and slung a coffee pot over it. They then built a small fire; and when the coffee was boiling, they hung some pots of chicken and beef pie to warm. The ladies meanwhile attended to the vinegar salad and corn biscuits they had set up on the tailboards of the wagons, and Willacassa joined them.

When the clambed had burnt itself out, the live coals were swept aside with a small cedar branch and buried in wet sand. Then it was everybody for himself. Pulling the clamshells apart—when they didn't fall apart on their own—diners impaled the meat of the clams on small, sharp forks and filled their plates. They piled the same plates high with biscuits and vinegar salad taken from the tailboards of the wagons. Later they helped themselves to the chicken and beef pies and coffee. Having stocked themselves against any chance of recurring hunger, the members of the clambake dispersed to find places on the beach where they could sit with their legs folded and partake of their fixings.

During this whole ritual, John Grandee kept up a patter of pronouncements in his best Carolina accent.

"My, but this is good eatin'. Reminds me of my summers on the Sea Islands. Of course, I was a mere boy then. But I had the appetites of a man. Did you have your share of clams, honey?" This last was directed at his Creole mistress who had been singularly silent during the entire excursion.

"Don't worry about me, John. I can look after myself."

"I know what'll loosen you up. A bottle of beaujolais, which Mr. Harrah was good enough to give me out of his private stock."

But the girl would not be appeased, and she picked herself up to sit apart from him.

"Are you having a good time?" asked Francis of Willacassa, who had found a spot near the breakwaters.

"The clams are very tasty," she said. "I do hope your parents don't mind that we're sitting off by ourselves."

"They have Uncle John to entertain them, and the friends they brought along from the Mount Vernon."

"What's wrong with the Creole girl?"

"Lucienne?"

"Yes, she seems unusually moody. More so than usual."

"It's her way."

"Her way?"

"It's her way of telling Uncle John that she wants a new trinket. Only when she's dissatisfied does he buy her anything."

"You seem to understand the girl quite well."

"Women aren't all that different, are they?"

Willacassa bridled at the idea.

"To doltish men who like to think so, no!"

"You're angry with me again?"

"Shouldn't I be?" she snapped.

"No," said Francis. "I meant no harm. I said something I shouldn't have said. I'm sorry."

Willacassa was softened by the apology, and somewhat mollified by the sense of satisfaction the baked clams gave her.

When they had finished eating, Francis suggested a stroll on the strand.

"It's a perfect time for walking," he said. "The wind's down and the sky's bright."

"Aren't the wagons leaving soon?"

"Father's is. But the other might hang about."

"And if it doesn't?"

"Then we can walk back. It's not that far."

"They won't miss us?"

"I've already explained to Uncle John. He understands."

"What does he understand?" asked Willacassa obliquely. "That women are all the same? I'm not like that Creole girl."

"Lucienne?"

"Lucienne."

"I know that."

He took her arm, but ever so gently, only to guide her in the direction of their walk.

Willacassa acquiesced, but just barely.

"It's a glorious day," he said, pointing to a deep red sun and streaks of cloud stained with purple. Willacassa could not help but nod in agreement. The ocean too was shot with color, rolling coral pearls on a coral beach.

"It's a lover's day," muttered Francis, almost in a whisper.

Willacassa pretended not to hear him. "See that," she said, pointing up ahead. "It's an old shipwreck. They say it comes from the Orient. You should have seen how quickly the cargo was carted away. Some of it is still sitting in Cape Island parlors."

"Looks like a fish picked clean."

"Well, nothing's wasted on Cape Island—except perhaps time."

"Willacassa, don't you care for me at all?" Francis was determined to get their talk back on course.

"I wouldn't be here if I didn't."

He took her hand. "Then why are you so unromantic?"

"I don't like summer flirtations."

"Is that what you think this is?"

"I don't know what to think. I suspect you want someone like Lucienne. In fact, I'm not altogether sure you haven't had your way with her."

"Lucienne? She's Uncle John's girl."

"At least he seems to think so. I'm not so sure she's of the same mind."

"I don't want to talk about Lucienne. I want to talk about you, about us."

"We'll talk another day," said Willacassa.

"But it's so beautiful now. It seems a shame to waste such a setting."

And for a moment Willacassa agreed with him. It was as gorgeous a day as she had ever witnessed on Cape Island. The billowing clouds, the circling curlews, the endless, rocking, churning ocean. It made her dizzy to watch. And she knew it would not be the same tomorrow. Francis was right about that. But tomorrow was what counted if today counted for anything at all.

"Let's go back," she whispered. She did not however remove her hand from his.

HORSEBACK

"Do you ride?" asked Francis Grandee one evening when he had escorted her home after a band concert on the lawn of Congress Hall.

"A little." In fact, Willacassa had been taught to ride tomboy-style by a laughing Henry Sawyer when she was fourteen.

"Good. Then I'll have a horse for you at sunrise and we'll ride the length of the strand."

Willacassa was waiting for him when Francis Grandee in a muslin shirt on a hot August morning appeared on horseback at Trescott House, with the reins of another horse in hand.

"Whose trousers are you wearing!" His laugh was broad, spontaneous, almost gasping as he caught sight of her "riding clothes." With the ocean at his back, he cut a wildly romantic figure as a red sun began slashing across fugitive clouds.

"My brother's. If you'd rather ride with some fluff or whipped-cream cake, you can find any number of them at the Columbia House."

"It's just that you're prettier in a dress!" He laughed again in short bursts. "But—But pretty just the same. Here, let me get a saddle that would suit you."

"Never mind. My father has one in the stable."

She disappeared behind the house and in a few minutes, hauling with some difficulty, she returned with an English riding saddle.

Grandee dismounted and removed the sidesaddle from the dapple gray. Throwing her father's saddle over the horse, he proceeded to secure it. When it passed inspection, he helped Willacassa to mount.

"Race you to the seabreakers!" she called as she pointed her horse toward the strand.

Grandee accepted the challenge. And though Willacassa had a good headstart, he lost no time overtaking her.

Willacassa slipped out of her stirrups and jumped down. "You call this a horse! She must be seven years old! Trade mounts with me and I'll give you a run for your money!"

Gallantly Grandee obliged. This time he had to press hard to gain ground. Whipping from side to side, he managed to catch up with Willacassa just where the rescue boat was beached. As he sped past her, Willacassa's horse landed in a pool of soft sand and stumbled.

"Are you all right?" Grandee was all solicitude as he drew near. But Willacassa held her saddle and waved him off.

"Oh, this is nothing? You've got to see the sand at Higbee's Beach."

"Higbee's Beach?"

"Just north of the Landing."

"If it's as bad as you say," pronounced Grandee, "I'd like to see it."

"Not just now. Later. I haven't eaten yet."

"Then we'll take breakfast at the Landing House. I'm told it's open all hours."

"Have you been inside the Landing House?"

"No."

"Well, it's not like the Mount Vernon, Francis."

"Doesn't matter," he laughed. "We're not dressed for the Mount Vernon anyway."

"I believe you're making fun of me again," Willacassa declared. But she did not wait for a reply. Turning her horse about, she cut across the windswept dunes to pick up the road to the Landing House.

But the nearer Willacassa drew to Higbee's Beach, the more relunctant she was to get there. And Grandee had trouble figuring out why. At the Cape May Landing House she dawdled over her coffee and eggs, even taking the trouble to exchange pleasantries with its proprietor, John C. Little. Later she suggested they scour the pebbly beach for Cape May diamonds. "Just for a little while," she said.

"What about Higbee's Beach?"

"Higbee's Beach can wait."

They walked down to the bay's edge and hunched over the sand, looking for smooth stones the water had washed up.

"Here's a perfect one," announced Willacassa, holding up a pure crystal of blanched quartz, the size of a penny.

"What about these?" asked Grandee, cupping a handful of shiny pebbles along with some bay water.

"Very good. You've got a topaz and a jasper there." She singled them out and added them to her find. "But the rest are misshapen or imperfect."

"Pretty nonetheless," remarked Grandee. He inspected the next stretch of sand. "Here's something!" he called, reaching down. A moment later he exclaimed, "Oh, my God!"

Willacassa looked up, then broke into ripples of laughter as Grandee rid himself of the jellyfish he had plucked from the water, hurling it fifty feet along the shore.

"Anyone can tell you're not a year-rounder!"

"Let's go on to Higbee's Beach," he insisted.

"All right." But there was a note of reluctance in Willacassa's voice. Returning to their horses, they rode north along the quiet surf.

"It must be near noon. The sun's directly above us. Is that the beach up ahead?" Grandee pointed in the distance to a stretch of sand and dunes and dwarf pines that was of a decidedly different character from the beach at the Landing.

Willacassa nodded, bringing her horse to a halt.

"What's the matter?" he asked.

"Nothing. Only you'd better go on without me."

"I wouldn't think of it."

"I really don't like Higbee's Beach," she said.

What he saw in the distance struck Grandee as a lonely, desolate place, overgrown with vegetation and dwarfed trees. But he did not see it as forbidding, leastwise near the water.

"I'll not take advantage of you there," he said. "If that's what you're afraid of."

Willacassa did not break into the smile he had expected.

"It's not that."

"What then?"

"I don't know. Phantoms of the past, I suppose."

Grandee could see that she was quite serious, her color all but gone, her eyes almost vacant.

"What phantoms?"

"A half dozen slaves drowned here some years ago. Only a hundred yards from where we stand."

"Slaves? Why should that bother you?"

"They were thrown overboard."

"To avoid detection?"

"Yes. I saw the ship when it was wrecked and found the slaves washed ashore a few days later."

"Still, that's no reason to—"

"I'm not going with you, Francis. I'll wait till you return."

"Then I won't go either."

"No, I want you to go. You came all this way to see Higbee's Beach. And I want you to see it!"

Grandee did not know what to make of Willacassa's sudden fear. But he decided to humor her.

"All right. I won't be long."

He followed the beach that stretched up ahead, moving his horse slowly and deliberately through the soft, yielding sand. Though the sun was strong, Grandee felt the wind off the bay and the vapors of the dying vegetation. This mixture of dry and moist air settled on his skin, rendering it chilled and clammy. It was not unlike the feeling he had when he was overseeing the work of Negroes in the marshes of his plantation.

With the shift of wind to a land breeze, Grandee began to dry off. Cautiously he resumed his exploration of the place. He did not know where the slaves had washed up on shore. Nor did he much care. He was used to blacks, the quick and the dead. Evidently Willacassa was not. She was probably a child when it happened, and the sight of half-a-dozen dead black men sprawled out on the beach must have frightened her, over the years haunted her. He could see she was no longer rational on this score.

But then again, thought Grandee, few Northerners were rational on the subject of slavery. They saw slavery as some sort of hobgoblin. They did not understand that Southern gentlemen were their black brothers' keepers. Without white overseers to look after them, the blacks would starve or run wild and ultimately perish. Slavery was mutually advantageous. And an institution that was mutually advantageous could not be all bad.

He returned to Willacassa prepared to put into words what had just run through his mind. But when he saw her sitting silently on her horse, her girlish figure silhouetted against a sun-drenched sky, he thought better of the idea. The time was for reassurance, not words. Instinctively he knew this. Talk could be death to romance, and the mood of the moment was decidedly romantic.

DOUBLE PORTRAIT

During the last week of his stay on Cape Island, Francis Grandee saw Willacassa daily. They had ice cream together or dinner at the Landing or visited the towering, ninety-four foot lighthouse on Great Island. Near the end of one of these strolls they came across a group of vacationers posing for a photograph in front of "McMakin's Atlantic Hotel," the letters boldly spelling out the name between rows of windows at the top storey of the popular watering place.

"I like that," said Francis.

"What?" asked Willacassa.

"The idea of a photograph. Why don't we pose for a picture too?"

"Here?"

"Or in the man's studio—if he has one."

As soon as the picture of the vacationers was taken and the crowd dispersed, most of them heading for the dining hall as the dinner bell rang, Francis approached the photographer.

"Can you take a picture of the two of us?"

"I have some other buildings to photograph." The man was getting his paraphernalia together and loading it on his wagon.

"When you're through?"

"I work for Langenheim's of Philadelphia. If you want a daguerrotype, I can do your portrait. I've taken a room over J. M. Smith's Clothing Store for a few weeks. Come in tomorrow afternoon."

"Do you know where that is?" asked Francis of Willacassa.

"Yes, I do."

But Willacassa was not sure she wanted the picture taken. A double portrait was a kind of commitment, rather like an engagement. After all, they would be seen together forever after. And no whisper of a betrothal had ever been made.

When Francis called for her the next afternoon, she was given no time to reconsider. Before she could catch her breath she was whisked into a carriage and a few minutes later was sitting with him in front of a camera.

"Hold still!"

They did. Then in a blaze of light and a puff of smoke the thing was done."

"I shall keep it with me always," said Francis when they picked up the portrait a couple of days later.

"I'd rather I kept it," said Willacassa.

"To remember me by?"

"Something like that."

"Then it's yours."

When she was alone, Willacassa studied the portrait. They were a handsome young couple, anyone could see that. She looked at herself sitting on a regal chair and Francis standing just behind her. And each time she thought she was finished with the picture, she looked at it again. But she did not show the portrait to her father. Instead she hid the picture case away. And she would continue to keep it hidden until such time as it would seem proper to bring it out again.

THE END OF SUMMER

The day of departure arrived. A mass evacuation was already taking place and "closed" signs appeared in the windows of many of the stores on Cape Island. The long summer was over, and the merchants were anxious to take a holiday of their own. The turnpike was cluttered with carriages and wagons piled high with baggage. Arguments arose as to who was at fault for the traffic jam. Departing visitors were afraid of missing their steamboats, and a few dilatory individuals were scolded by their wives for not having purchased tickets in advance.

The Grandees were almost the last to leave. All morning Willacassa had expected Francis to stop at Trescott House to bid her goodbye. But it was not until mid-afternoon that he appeared and asked for Willacassa and Harrah. He was told by Rachel that Mr. Harrah had taken some of his guests to the Landing but that Willacassa would be down shortly.

"I thought you had gone by now," said Willacassa as she caught sight of him in the entryway.

"I wouldn't go without seeing you."

"I had hoped not."

"It was a glorious summer," Francis began.

"It was that."

"I can't thank you enough for being such a good companion."

"Thanks are not necessary, Francis."

"You know what I mean."

Willacassa nodded, growing impatient with him.

"Well, then, it's goodbye for a while," said young Grandee.

"Yes." Somehow Willacassa never expected that her summer with Francis would ever end, leastwise end as a summer flirtation. Other romances ended that way, but hers she believed would be different; after all, she was different.

"I'll write," he promised.

"It's too bad you don't live in Philadelphia," said Willacassa.

"Philadelphia?"

"Yes. Then you could telegraph messages to me via the Philadelphia and Cape Island Telegraph Company."

Grandee detected the edge of sarcasm in her voice, but pretended not to notice.

"Maybe they'll put a line up between Savannah and Cape Island. Do you think they might do that for us?"

"If you're going, you'd better go," said Willacassa. "Your folks will be waiting for you. Your uncle John has already left for the steamer.''

"I just wanted to say—''

"That we had an exhilarating time together?''

"That I adore you.''

This Willacassa was not prepared for. She had half formulated a reply to whatever innocuous statement Francis might make. But his declaration changed all that.

"You heard me, didn't you, Willacassa?''

Willacassa was deliciously dumbstruck.

"What I mean is that I love you. I wanted to tell you before today, but I was afraid it would spoil things between us.''

"Spoil things?''

"I didn't want you to think it was a summer flirtation, or anything like that. I fully expect to be back on Cape Island next year. There's no place I'd rather be.''

He waited for a reply but none was forthcoming.

"Well, aren't you going to say something?''

Willacassa broke into a warm smile. "I'll write too," she said. "I may even do a sketch or two for you. Don't want you to forget what the place is like.''

"And you? How do you feel about me?'' he asked.

"That's not something a Southern gentleman should ask a young lady. Either he knows or he doesn't know. I'm surprised at you, Francis Grandee.''

"Perhaps you're right, Willacassa.''

"Of course I'm right. Living south of the Mason-Dixon line has given me a special insight on the subject," she teased. "What will you be doing this winter?''

"Overseeing the plantation. Training with my regiment. Thinking of you. What else can a Southern gentleman do?''

"He can re-establish friendships with other young ladies.''

"If they look like you and talk like you and frazzle my heart as you do, perhaps I will." He hesitated a moment and then added. "But it's not likely.''

"I heard that Southern girls are quite pretty, even prettier than I am.''

"They are pretty. And they come from good families. But no one of them can ride a horse like you. Or put a 'hop' to shame like you. And none of them are quite as pretty as you—at least as I see it.''

He looked tenuously at her once more.

"I think I've said enough. And if I don't do it now, I guess I never will.''

So saying, he kissed her, first on the cheek and then on the lips. Pulling himself away, Francis made directly for his carriage. "Say goodbye to your father for me!" Once he was aboard, he wheeled his carriage about and waved a final farewell.

Willacassa stood speechless. As she watched young Grandee drive spiritedly away, she did not move from the spot.

CONFLAGRATION

The Grandees, in fact all of the summer visitors, had been gone almost a week when Willacassa rose to the clatter of bells and steamboat whistles and saw from her window a harsh red glow in the sky. She rushed downstairs to awaken her father but Harrah was already dressed and into the street, cinching his saddle.

"Where is it?" she called from the doorway.

"Budd's Island," shouted Harrah, alluding to that tract of land by its old name. "It's the Mount Vernon." He mounted his horse and sped away.

By the time he reached the hotel, the entire front of the gigantic structure facing north toward the turnpike was engulfed in torrents of flame, shooting dark billows of smoke and flambeaus of debris into the night sky. From the windows of the towers in the middle of the hotel and at each end, fire and smoke poured forth in puffs of black and red rage. A few men were seen running back and forth, but there was no organized brigade and no hope of getting the fire under control.

Harrah caught sight of Henry Sawyer coming towards him with a bucket of water.

"Here, throw this over me," he shouted. 'I'm going inside to look around." But even soaking wet he could get no nearer than the open door for all the heat and blistering smoke.

"There can't be any guests in there," called Harrah. "They're all gone by now."

"It's the Cains I'm worried about. No one has seen them."

"The Cains?"

By now the fire was consuming the east wing, ravaging the great dining hall, and shooting its way along the walls and up the high towers. Someone waved the awe-struck bystanders back.

"The walls can't stay up much longer! Get back! Move away!"

Henry Sawyer grabbed Harrah's arm and the two men moved out of range.

"There's no hope for it," said Sawyer. "It'll come tumbling down. The world's largest hotel, nothing but a pile of rubble."

And as they stood and watched the growing inferno burn itself out, casting startling flashes of light and shadow on everything in its vicinity and standing as a beacon for miles around, both men realized that Cape Island would never be the same.

A wiry Joseph Leach, wearing trousers over his nightshirt, appeared, notebook in hand. And behind him, standing in his carriage with a pad of her own in hand, watched Willacassa. Leach's bony face caught the flicking red and yellow incandescence of the intense heat and brilliant fire. The publisher of the *Ocean Wave* kept shaking his head in disbelief

"How can something so magnificent, so elegant, just vanish before our eyes?"

And for almost an hour the onlookers watched until the once glorious hotel, the darling of the *London Illustrated News*, was nothing but a smoldering ruin.

Then someone made the discovery that Philip Cain's namesake, though badly burned, had succeeded in escaping the blazing building, only to die the next day. But his father, one of the pro-prietors of the Mount Vernon, a brother, two sisters, and the housekeeper Mrs. Albertson, the only other occupants of the hotel, had already perished.

Joseph Leach's editorial that week said it best.

"Cape Island is no stranger to fire. Isaac M. Church, first Mayor of our incorporated city, urged us in his inaugural address to launch a program of fire inspection and fire prevention and the forma-tion of a hook and ladder company. But there was no way the Mount Vernon Hotel could have been saved or the lives spared of the six who perished in the fire. The truth is, beautiful as it was, with all its architectural magnificence, the Mount Vernon Hotel was too big. In everything there is a point of diminishing returns.

"Our great nation is growing fast. (I won't mention the divisiveness that is growing with it.) Cape Island itself has ex-perienced astonishing growth in the past ten years, not only as a watering place, but also as a meeting ground of North and South. But there reaches a point when size itself can be its own encum-brance. It may be that in the matter of resort hotels smaller is bet-ter. The once mighty Mount Vernon Hotel had three-and-a-half times the capacity of its largest competitor, the Columbia House. And the building of its projected west wing had not yet even begun! How many servants it employed is still a matter of conjecture, though there is no doubt it fostered a whole servant class. We are extremely fortunate that what happened on September 6 did not occur at the height of the summer season.

"To see the Mount Vernon Hotel go up in flames was a shat-tering experience for all who managed to make it to the site. No one who saw the conflagration will ever forget it. (Please note Willacassa Harrah's line drawing of the event on page four.) Perhaps there is a lesson to be learned here, albeit a painful and a tragic one. Let us pray it is not a foreshadowing of conflagrations yet to come."

1857

BETROTHAL

BY MAIL

Willacassa and Francis Grandee became engaged by mail. Francis Grandee wrote a flowery prose filled with poetic allusions and borrowings from Richard Lovelace, Sir Charles Sedley, and Robert Herrick, which he had copied out of a book in his father's library.

"If to be absent were to be
    Away from thee;
Or that when I am gone,
You or I were alone;
Then, my Willacassa, might I crave
Pity from blustering wind, or swallowing wave."

"But I am tied to very thee
    By every thought I have,
Thy face I only care to see,
    Thy Heart I only crave.

All that in Woman is adored
    In thy dear self I find."

"Thou art my life, my love, my heart,
    The very eyes of me:
And hast command of every part,
    To live and die for thee."

Willacassa read all this with a mixture of amusement, embarrassment, and a searing passion for Francis Grandee. For behind the words and borrowed passages, she saw the handsome face of the young man she had come to love. And though she found much to mock and good-naturedly make fun of in Francis Grandee, she came even in his letters under the spell of his youthful, dashing exuberance.

And when that cold day in March, with the winds howling outside the windows of Trescott House and the waves tossing and rolling recklessly offshore, she read his letter of proposal, she did not wait to finish.

"Father, Daddy!" she came running. "Guess what! Francis has proposed! Here, read this!"

Harrah put on his glasses. The letter contained a proposal all right, in that oblique Southern way of young Grandee.

"And when I come to see you in April, I will carry in my

heart—and in my pocket—a message of undying love. I've already told my parents and they are pleased as pickling juice. All we have to do is set the date.

"But first you will visit us at home and see what kind of family and what nature of life you will be marrying into. South Carolina—at least our part of the woods—is the nearest thing to a heaven on earth. It has its own pace, its own charm, its special beauty. And all that will be enhanced by your lovely presence.

"I can't wait to see you.

All my love,
Francis"

Harrah put the letter down and removed his glasses.

"Well, what do you think?" asked his daughter.

"He's not giving you a chance to say 'Yes,' Willacassa. He's saying it for you."

"You know I'd say 'Yes.' ' "

"Still, he should give you that option."

"Are you serious, Daddy?"

"Not really. I was only commenting on a matter of form, not substance."

"Then you approve?" she smiled.

"Of course I approve."

But did he? Harrah asked himself in private. Did he want Willacassa living in the South, with all that living there implied in these times? And they were troubled times indeed. Blood was being spilled in Kansas. A vacillating James Buchanan had been elected to the White House to succeed an ineffectual Franklin Pierce. And the Dred Scott Decision had all but guaranteed permanent servitude for the Negro, throwing the Missouri Compromise out the window and threatening to spread slavery in the territories.

And then there was his own personal decision to reflect on, the decision he had made several years ago. At great pain he had allowed Madeleine Culpepper to get away from him. Ostensibly he had done this because he did not want his children to grow up on a Virginia plantation, white masters to black slaves. but Willacassa was now prepared to begin a new life in the deep South. And Jonathan was gone—who knew where? The boy had never written. All that Harrah had sacrificed for, did it no longer hold as reasonable or viable? Except for his own searing dislike of the feudal system that the Southern aristocracy was ready to fight for, even die for, his hard decision now appeared an empty sacrifice.

And it stood as just another of the ironies that made his life such a puzzlement to him, a life that had given him Cassandra Trescott and then snatched her away, a life that had teased him with Madeleine Culpepper and then cut him off from any chance of marrying her.

But, of course, there was something more that disturbed Harrah. In losing Willacassa, he was losing all the family he had. It would make little difference if she were moving on to the Oregon Territory or California instead of South Carolina. They would still be miles and worlds apart. He had always known that she would marry some day, but he had hoped, with good reason, that she would remain in the Cape May area. Now that hope was shattered.

HOUSE GUEST

From the moment Willacassa saw the Grandee plantation, she had an uneasy feeling about the place. It was, of course, much like the picture books she had seen of the South. Wide, sweeping driveways in a park-like setting. Avenues of trees and hanging moss stretching like a carpet. Servants and working people walking the grounds, if not exactly attired in their Sunday best, reasonably clean in their flaxen shirts and dresses. And on a rising sward, the big house, white and masterful, its Grecian pillars gleaming ivory in the sun.

But even with a solicitous Francis as her guide, taking every opportunity to seize her arm and call her attention to something she had missed—the ice house, the new barn, the sawmill—she found the atmosphere as oppressive as it was lush, forbidding as the road to the plantation was marshy.

There was a buzz in the air, a drone, which she attributed to the insects and the everpresent whirring of birds on the wing, that all but hypnotized her with their sameness. And she longed to hear the boom of the ocean, to have gusts of sea air blow free again. But she knew that this could not be, and that for the next week or two she would remain a guest of the Grandees and that despite her desire to be with Francis and to become acquainted with his family she would find being in the big house terribly confining.

The house itself was large and spacious with countless floor-to-ceiling windows. The main room impressed Willacassa as an oversized, carpeted widow's walk except that there was no sense of air at all and there was very little light trickling in from the windows, only Northern light. The chandeliers, which sparkled and cast a great brilliance of light during the evening hours, hung heavy and unused during the day. But anyone could see that the place was expensively furnished, Empire in mode, with lovely carved statuary and curved side pieces with brass and ivory ornamentation.

Still, there was something about the Big House that showed neglect. For one thing, the curtains were always drawn, making for a musty and dank odor in the rooms. And though there were house servants aplenty, moving silently about the place, every so often flicking a feather mop in their wake, a fine coating of "country dust" covered everything, even the dinner plates which were stacked clean the night before. Willacassa's years at Trescott House enabled her to see these things, she realized, where others might not have noticed. But they made it no easier for her to dismiss them as inconsequential.

In contrast to this, the Grandees were always decked out in the cleanest linen and they smelled washed and soaped and perfumed. Even Francis had this fragrance, though his was a subtle scent smelling by turns of English leather—his boots perhaps—or sweet tobacco. Yet he smoked segars when he did smoke, not a pipe.

Still, it was with some distaste that Willacassa sat down to supper, feeling a clamminess as she eyed the dinnerware and touched the table's edge. But once she partook of the cooking, she found the food to be exotic and tasteful and she forgot about the condition of the plates. Seafood was the exception rather than the rule here. And though Willacassa missed her oysters and steamed clams and flounder and drumfish, she developed an absolute passion for Southern-style chicken.

But that was as far as her passion went. Except for Francis, she was not all that fond of the Grandees. Oh, his mother was gracious enough and well-spoken. Of all the Grandees, she was the most cultured and the least grating. She was familiar with the poetry of Byron and had read the novels of Sir Walter Scott and Jane Austen. Of course, she was not really a Grandee. She was a Simpson; and if Willacassa remembered correctly, the daughter of a Congressman from Louisiana.

But John Grandee, Francis' oversized uncle, was an absolute lecher. He had no sooner left the bed of his Creole mistress than he was shopping about again for a Negro girl to pinch or one of his own kind to titillate with an off-color joke. Somehow he did not have the nerve to treat Willacassa in this way or else by the time he caught her attention he had run out of stories.

But it was Hazon Grandee, Francis' father, that she liked least of all. Hazon Grandee talked like no one she had ever heard before. He referred to his slaves as halfway between mules and horses. And he likened the abolitionists to rabid dogs.

"They and the clergy are to blame for the troubles in this country. They are, you see, of the Puritan race. The Cavaliers and Jacobites and Huguenots who settled the South naturally hate and condemn and despise the Puritans who settled the North. The Southerners come from master races. The Puritans a slave race, descendants of Saxon serfs. Southerners are basically of Mediterranean races, descendants of the Romans. Our Cavaliers and Jacobites are of Norman descent, as are the Huguenots. The Saxons and Angles, ancestors of the Yankees, came from the cold and marshy regions of the North, where man is little more than a cold-blooded, amphibious biped."

Willacassa thought he would at least spare Northern women,

but he did not.

"Our women are all conservatives—moral, religious, and sensitively modest. They abhor Northern women for infidelity, immorality, and gross licentiousness."

"But Father," protested Francis.

"I'm not referring to Willacassa, my boy. I don't regard her as a Northerner. Cape Island, after all, is south of the Mason-Dixon line."

If Hazon Grandee thought this would mollify Willacassa, he was grossly mistaken.

"I never heard such vituperation," protested Willacassa when she was alone with Francis once more.

"He meant no harm."

"What did he mean, then, Francis?"

"He's only expressing deep conviction."

"Deep conviction? The lower depths, you mean. And I suppose you share his sentiments?"

"Some of them. But I don't feel so strongly about it."

"How do you feel?"

"I feel we've gone far enough in this matter. Let's talk about something pleasant. Our engagement, for instance."

"I'm not sure I want to talk about that just now." Willacassa could not transform herself into the smitten, loving, lily-white damsel he wanted her to be at this moment.

"All right, then, let's talk about tomorrow."

"Tomorrow?"

"Yes, I'm going to pick up a new horse. I'd like you to come with me."

"A new horse?" Was that all on his mind? Couldn't he see how upset she was?

"I'm sorry. I don't want to go."

"I'll go myself and meet you afterwards. Maybe you'll be in a better frame of mind."

But Willacassa was not in better spirits when he showed off his new mount. In fact, she struggled all day just to be civil. She did not like herself for this, but she could not help it. She dreaded any encounter with Hazon Grandee and persuaded Francis to go for long walks just to avoid being near his family. Though she took dinner with the Grandees that evening, she found excuses to retire early. And despite Francis' gentle protestations, she made her way to her room, finding a sense of relief in locking the door behind her.

THE UNEXPECTED

The exertions of the trip and the past few days suddenly caught up with Willacassa. No sooner did she put her head down than she was ready for sleep. But a soft knock on the door of the adjoining room stirred her. She rose, slipped on a dressing gown once she found the sleeve, and, leaning as close to the door as possible, whispered, "Who is it?"

"It's Francis, Willacassa. Unlock the door."

"Unlock it?"

"Yes, turn the key twice and the door will open."

Willacassa did as instructed but was none too happy about doing it.

"I didn't know you were in the next room. I thought you were at the other end of the hall."

"I am. But I thought it would be more discreet to visit this way."

"Discreet? What are you talking about?"

"If someone saw me entering the room directly, he might think the worst."

"Well, I hardly think what you've done is any better. You have no business being here, Francis."

"Of course I have. I love you. Isn't that enough?"

"That's beside the point."

"And we're engaged, aren't we?"

"Yes, but engaged isn't married. There's time enough for this when we've taken our vows."

"I can take my vow right now, if it comes to that." He leaned over to kiss her.

Willacassa was warmed by the kiss and not without desire, but she felt put upon, advantage taken of her.

"All right, you've had your kiss. Now go!"

"You're not serious, are you?" He drew nearer to her and kissed her once more. "Tell me you're not serious."

Willacassa gently pushed him away.

"Don't spoil things, Francis. I like that you asked me to your home. And I am proud to be engaged to you. But this room is still mine, not ours. And I want to keep it that way for now."

"You sound like a Northern preacher's daughter," he said testily.

"I sound no different from your Southern girls, I believe. But even if I do, I mean what I say."

Francis retreated for a moment, then sat down on his fiancee's bed.

"I've waited a long time for you, Willacassa. Since last summer all I wanted was to take you in my arms."

"You have, Francis."

"And more." He lay back on her bed with his hands under his head. "Much more."

"Drive it out of your mind, Francis. I'll make love to you only when I'm ready. And I won't be ready until the wedding ceremony is over."

Francis lay back looking at her for some time, as if he would never really comprehend her. Then with a shrug of his shoulders, he returned to a sitting position. "Here, come sit beside me. I want to talk to you." He read the distress in her eyes. "Only talk, Willacassa."

With some misgivings she sat down beside him.

"Don't you know," he began, "that most engaged couples become sexually acquainted before marriage."

"I know nothing of that."

"Well, it's true. And no one cares one way or the other. It's taken for granted."

"It makes no difference, Francis. I have certain expectations about our life together and I will not be disappointed."

"Expectations? What kind of expectations?"

"I don't care to talk about it now."

She started to rise but Francis took her by the arm and slowly pulled her back down on the bed. At first Willacassa thought he meant only to return her to a sitting position. But she soon discovered that he wanted to pull her down on top of him and in doing so had put his hand on her breast.

"Francis, are you daft! Let go of me this instant!"

When he did not and instead pressed his lips against hers as much to keep her quiet as to satisfy his wants, she clenched her fists and began pummeling his chest.

Francis accepted the pummeling good-naturedly. But when Willacassa pulled back and struck him across the face just under the eye, he lost his composure.

"Why did you do that!"

"I told you to let go of me!"

At last he released her.

"Now leave!" She almost choked the words out. "I'm so angry I could cry!"

"Cry? Damn it, what for! You'd think I was trying to rape you!" He left the bed and stood as tall as he could make himself. "I won't have any more of this," he said. "If you're so bent on preserving your virginity, I'll look elsewhere!'' He turned to face her. "Did you hear me?''

"I'm so ashamed, Francis. I don't know what's gotten into you. All I know is I'm humiliated—for you, for me. What if someone heard us?''

"Who the hell cares!'' And with that he stormed out of the room, using the door he had entered by.

ACCUSATIONS

The encounter left Willacassa in a state of self-doubt. Though she was outraged by Francis' uncouth behavior and more than a little irritated by his threat to look elsewhere, she wondered whether in some way she were to blame for his actions. Was there a tacit understanding that engaged couples experiment with sex, in fact indulge in a foretaste of things to come? Willacassa did not know. Nor did she much care. It was not her understanding of things, and she would deal with the situation as she saw fit, not as Francis interpreted it.

At the same time, she herself was not without desire for Francis—nothing so gross as he had in mind—but desire nonetheless. And she wondered whether he sensed this and whether she could absolve him from blame because of his sensitivity to her feelings. Willacassa spent a restless night tossing these ideas about in the misty darkness of her room and when morning slowly stole through the slats of the shutters she managed to doze off for a while.

She took breakfast with the family at the big table on the lawn, for she wanted no one to think anything untoward had happened. She even sat in the chair next to Francis who barely nodded a morning greeting. She listened dutifully as Uncle John recited the latest news from the morning paper, and she nodded as each member of the family left the table to return to the house.

When they were alone, Willacassa spoke to Francis but carefully avoided meeting his glance.

"I'm sorry about last night,"she said. "But I just couldn't—"

"No need to talk about it."

"I hope you understand. It was a nightmare for me. It just wasn't like you."

"I'm sorry too. I didn't mean to offend you."

"Then you're not angry with me?"

"Angry? No. Just disappointed."

Willacassa lowered her head. "I'm disappointed too. I never expected you to behave in such a way. You were so unreasonable, Francis. So self-indulgent. You were not the person I knew on Cape Island."

"I'm the same person."

"I like to think so."

Francis turned his gaze directly on her.

"Will it be any different tonight?"

Willacassa's neck turned scarlet.

"Of course not! Don't tell me you still haven't given up on the idea. What's gotten into you, Francis?"

"But I love you, Willacassa!"

"If you love me, you'll wait. Like the Southern gentleman you are."

A knowing smile crossed his face. "I'm afraid," he muttered, "you've got a lot to learn about the South."

Just then they were interrupted by a disturbance in the big house. Willacassa heard what appeared to be a quarrel between Hazon Grandee and his wife. Voices were shrill and rising.

Seconds later Mrs. Grandee appeared at the door and signalled to her son.

"May I see you for a moment, Francis?"

"Not now, Mother. Willacassa and I are having a talk."

His father appeared at the door. "Get the hell over here, Francis! And be quick about it."

Hazon Grandee was not one to trifle with, especially when he was angry. Still, Francis held back, reluctant to deal with his father just now.

"You'd better go," advised Willacassa.

As Francis started towards the house, Willacassa wondered whether the dispute concerned her in any way. Her first reaction was to think so. But something in Mrs. Grandee's tone had suggested otherwise.

Willacassa stood alone for a few minutes. Feeling uncomfortable in the role of odd person out, she started towards the fountain to watch the water dribble the time away.

She had not gone far when Mrs. Grandee called her name.

"Willacassa, can we see you for a moment, darling?"

Despite Mrs. Grandee's everpresent gentility and the appellation, "Darling," Willacassa sensed a strain in her voice that could not be smoothed over.

"Of course, Mrs. Grandee."

When she had entered the house, she found his parents, Francis, and two Negro women standing in the main drawing room, with the younger of the two women weeping violently.

"Ah'm not sayin' young Francis ain't a fine mastah," the older Negress declared. "And Ah'm not sayin' boys won't be boys. It's a priv'l'ge survin' the Grandees. But Melva here is fo' months gone. She's been made pregnant by Noah, Noah who'll be her man. 'Taint right that someone come foolin' with her while she be pregnant."

"I told you, woman, that I had nothin' to do with her," an indignant Francis protested while his mother and father looked on. "I only bother with my own kind."

"Now, that ain't true, young mastah. You bother'd Melva before. And a couple others I could name. The Bible says you must be truthful."

"What do you know about the Bible, Mother Lou? You're ignorant. You never learned to read."

"Ah knows my Bible from Prayer Meetin's. Ah take it to heart. And Ah knows you went to Melva's cabin las' night and had yo' fill a her."

Francis laughed at the idea and looked to his mother and father for approval. But they remained stone-faced.

"I was in my room all night."

"I tell you, Mastah," Mother Lou said, turning to Hazon Grandee, "Ah saw him leavin' the cabin."

Hazon Grandee turned a wrathful scowl towards his son. "Mother Lou don't lie, Francis. I know her more than forty years. It's not that you took Melva bothers me. It's that you're supposed to be engaged to Willacassa. And she a guest in our house. What the hell is goin' on?"

Willacassa was dumbstruck listening to all this. She could not bear to look at Francis or the distraught Negro girl. She couldn't even tell whether the girl was pretty or not. More than ever since her arrival, she saw herself a stranger here. So long as this remained a family quarrel she had been able somehow to contain her searing humiliation. But now that Hazon Grandee had dragged her name into the dispute and she had become a party to the sordid business, she wanted to be gone from the place at once.

"As I said before, I was not with Melva. I was with Willacassa last night. I couldn't be in two places at one time."

"Do you know what you are saying, Francis?" his mother interjected.

"Watch your tongue, Francis," his father cautioned him. But in the same breath he turned to Willacassa. "Is that true? Is what he's saying true?"

"I can't believe what I'm listening to," said Willacassa, putting her hands to her ears.

"You'd better go now," said Hazon Grandee to the two Negro women. "I'll talk to you later. Right now we've got family business to settle."

When the black women were gone, Hazon turned to Willacassa. "I apologize for my son. We're none of us saints. But there's no excuse for his behavior last night—or today. What occurred between you two is none of our affair."

"Nothing occurred between us, Mr. Grandee. You'll have to believe that. And I can assure you nothing ever will!" So saying,

Willacassa turned on her heel and went upstairs.

Mrs. Grandee joined her in her room. She watched quietly as Willacassa threw her things together and started to pack them into a bag.

"I know it was a terrible moment for you downstairs, and I'm sorry, Dear. I only asked you to witness it because Francis had said he was with you last night. I thought Mother Lou and Melva were lying."

"I don't care to talk about it."

"You're hurt. And I don't blame you. It was an ugly scene. No place for a sweet girl like you."

"I'm not sweet!"

Mrs. Grandee sat down on the bed.

"I like you, Willacassa. It made no difference to me that you were not a Southern girl. Or that you did not come from a distinguished family."

"If you mean 'money,' my father does all right."

"Of course he does. But I don't want you to have any bitterness towards us. And though I'm not blind to his faults, I like Francis, too. Still, you must understand. Things are different here from Cape Island. Yours is an open society. Ours is closed. And it has all the aberrations and evils of a closed society."

Willacassa stopped just long enough to show her displeasure.

"Even though I came from the South myself, I was shocked when I first arrived here. What passes for normal in plantation life is quite extraordinary to the outsider. But in time you get used to it. It's part of being an aristocracy."

"I'm getting used to nothing, Mrs. Grandee. It's bad enough you're slave and master here. Your aristocracy, as you call it. But taking advantage of these poor wretches—in that way—is sickening. You have no idea how disgusted I feel. If I never saw a plantation again, if I never saw Francis again, it wouldn't be too soon."

"Listen to me, Willacassa."

"Listen to what? To how glorious your 'peculiar institution' is? Yes. I've seen the phrase in the newspapers. It's peculiar, all right. Sickening would be a better word. Vile and disgusting would be appropriate too. All I know is I won't be a part of it!"

Mrs. Grandee gave up trying to mollify Willacassa. With a sad turn of the head, she rose to leave. Before she reached the door, she heard her name.

"Mrs. Grandee!" Willacassa blurted out. "I meant nothing personal in all this. I believe you have my best interests at heart. And I'm thankful for that. But if you could arrange a carriage for me, I'd be most grateful. I'd—I'd like to leave without delay."

Mrs. Grandee sadly acquiesced. A carriage was procured, and within an hour Willacassa was on it. At first only Mr. and Mrs. Grandee stood quietly by to see her off. Then at the last instant Francis appeared.

"I'm sorry, Willacassa. You must believe me. I know nothing's gone well since last night. I've acted abominably. I've done things I would never have done. But the idea of being so close to you and not being able to—"

"I don't want to talk about it!"

At this, Hazon Grandee and his wife retreated out of earshot.

"But we must talk about it." insisted Francis, almost in a whisper. "So much depends on it. I don't want you to think what I did was commonplace behavior for me. I went berserk. I was so much in love with you, I did not stop and think. I acted on impulse, terrible impulse. I wanted to get even. You must forgive me!"

Francis leaned forward as he said this. And his face wore an expression of incredible pain.

But Willacassa was not moved. Nothing could have moved her at that moment. She was gutted by the pain of truth. And the truth was Francis had taken advantage of a pathetically helpless young woman. And in the attempt to cover up what he had done, he had shamelessly compromised her before his family. All they had been to each other, all the love she had harbored for him, was nothing in the face of this.

"I'm sorry, Francis. I'm returning to Cape Island. And it would be best if I didn't see you again."

Young Grandee took a step backward. His boyish face was even more boyish in the grip of reprimand. But a Southern pride intervened, rescuing it and returning the face to manhood once more. And that was the face Willacassa last glimpsed when the horses started up and her carriage rolled mournfully along the path away from the big house which, as she looked back at it at the turn, now loomed as nothing more than a big white sepulchre.

AFTERMATH

During the long voyage home Willacassa was as wretched as she had ever been in her life. Until she had arranged passage in Savannah on a ship destined for Philadelphia, she had been totally numbed by what had taken place. When she tried to think about it, the experience was so painful that she simply drove all thoughts of the plantation and the Grandees out of her mind.

But once she was out on the open sea, the clutter of her mind broke free and rolled back and forth like loose cargo in a ship's hold. Try as she would, Willacassa could not contain her emotions. They pounded against the sides of her heart until she thought its walls would cave in.

As she stood leaning against the deck rail in the morning, a ship's officer stopped to inquire if she were all right.

"All right? Yes, I'm fine."

But not for a moment did she or the officer believe this.

Later someone noticed that she had not eaten since she came aboard.

"I'll not eat again," she said. "Least of all, not while aboard this ship."

"Is there something wrong?"

"Wrong? No, there's nothing wrong. Everything is as it should be."

"Then why not eat something?"

"Because if I starve my brain, I won't be able to think dreadful thoughts."

She was soon left alone.

The captain introduced himself.

"I understand you're from Cape Island."

Willacassa nodded.

"I don't regularly do this," he said. "But if you're not feeling well, I could drop you off there."

"Where?"

"Cape Island."

"But I thought we're sailing for Philadelphia."

"We are. And you can take a steamboat from there to Cape Island. But if you're not well, I could arrange to stop directly. Would you like that?"

Willacassa looked at the captain but without quite seeing him.

"Would I like that?"

Her emotional dam gave way and she broke into tears.

"Yes, I would like that very much."

Harrah knew his daughter well enough to sense something was wrong. He had expected she would be accompanied by Francis Grandee when she returned to Cape Island. When she came alone, there was no doubt in his mind that her trip to South Carolina had been a disaster. But he pretended ignorance.

"Willacassa! What a delightful surprise! How was your stay in the Southland?"

She kissed her father and held him close.

"It's so good to be home again. You've no idea. I'll never think ill of Cape Island again."

"You never did think ill of it. At least that I can remember."

"I did. I was an ingrate. But no more. I thank the stars I live in a free environment. The South, the South is like a dank, dark prison. The sun may shine there, but it shines only on a select few."

"Then it wasn't a happy trip?" Despite his daughter's obvious discomfort, Harrah was slyly pleased.

"Happy! It was the worst experience of my life. If I never go South again, I could die content."

"And Francis?"

"There is no Francis Grandee. Not for me! Do you understand?"

"It's that bad?"

"That bad."

Harrah had hoped to talk to his daughter about what had happened once she had settled in again and had a chance to rest. But after she had taken her dinner with Rachel—Rachel who maintained a discreet silence—she retired to her room.

It seemed that all Willacassa could do was sleep. She slept in her clothes all the rest of the day and the day after. And when she finally had enough of her bed, she kept to her room, even taking her meals there.

To Willacassa the world had not ended. The ships on the distant horizon were plying their trade. The beach, the ocean, and the sky were still in harmony with each other, intermingling tint and color. The sun was still pouring its rays on Cape Island. And the Columbia House across the street was as imposing as ever. Everything continued as it was; little had changed. But the light had gone out of Willacassa's life.

It was easy to say she was alive because her senses appeared to be intact. But her hearing was only humming faint, her taste buds flat, and there was no sense at all of wonder in what she saw. Everything stole up on her, then slipped away like a tide in ebb. There was no high water in her life, only low, and on her tongue the salty taste of gritty sand.

"Harriet Sawyer was here," said Harrah upon entering her room. "She'd like you to take dinner with her and Henry on Friday."

"I'd rather not see them."

As the Sawyers were favorites of Willacassa, Harrah felt he had played his trump card. But he tried again.

"Joseph Leach sent a note. A sketch artist from *Harper's Weekly* is visiting Cape Island when the summer season is under way. He thought you might want to meet him and see how he works. He'll drop in at the *Ocean Wave* office next Monday."

"Thank Mr. Leach. But I'll not be available."

"And what about me?"

"You, Father?"

"Yes, I'd like company for dinner. Will you be available for drum fish?"

Willacassa perked her lips in a dour smile.

"You know I can't say no to you, Daddy. I'll be there. Have Rachel call me when it's ready."

1858

AWAKENING

HARRIET

Willacassa should not have noticed the chambermaid her father had hired for the new season. For she was as ordinary and plain as the old brown boards which had survived the fire at the Mansion House. Harriet—why was it these colored women never had surnames?—was a short, stubby, square-faced woman who scarcely measured five feet tall. Though she was a good worker, she did not push herself too hard and fell asleep whenever she sat down to rest.

In the few instances when Willacassa had occasion to talk to her, she found that Harriet had several front teeth missing which affected her speech. But, though it was apparent the woman had no formal schooling, Harriet was by no means stupid or uninformed.

"Have you worked on Cape Island before?"

"Yes'M. I worked at the big hotels from time to time. Congress Hall and the Mansion House. Good thing I quit the Mansion House before the fire. Or I'd be without a job."

"How did you know to apply here?"

"Met your father at Mr. Edmunds's furniture place. I was fixin' torn seats on cane chairs till I could find a better job. It's hard on the fingers, you know. And he felt sorry for me. Done ask me if I wanted to work at Trescott House and I said Yes. I liked the man right away."

"Where are you from, Harriet?"

"From Maryland. But I like it better up here. Maryland's too South for me."

"Do you ever go back?"

"From time to time. My man's a slave there. I was workin' to buy his freedom. But he's took another wife. So now I jus' keep workin' for myself."

"Men are such animals!" complained Willacassa.

"Not all are like that. Anyway you should have no trouble, pretty as you are, Miss Harrah."

But the chambermaid's comments were of little comfort to Willacassa. The ache, the wrench of her break with Francis Grandee was still too painful for her to handle.

How did one explain what had happened to a stranger, to herself? Particularly to someone of Harriet's race? How did one deal with such a sordid business, mixed up as it was with girlishly mistaken notions of romance and emotional turmoil? She could find no solace anywhere, unless there was a solace in finding no solace.

"I probably have no right talking to you this way. But what's it like, Harriet? How does it feel to be a slave? To have no say over your own fate?"

The chambermaid suddenly became cautious. Did she know that Willacassa was overwrought? Did she not want to say anything that Willacassa in her calmer moments might interpret wrongly?

"How do you stand it, Harriet?"

"Stand what, Miss Harrah?"

"The insults, the humiliation."

"No one insults me here."

"I mean for your people. I die when I think of what it all means. Trouble is, I never really thought about it before. Except briefly, when I was a child and there was a shipwreck."

"A shipwreck?"

"Never mind." With a shake of her head, Willacassa dismissed the memory. "What I mean is, we lead sheltered lives here on Cape Island. Amid the grand houses and the 'hops' and the flirtations—we have no way of knowing what it's like down there."

"In the South?"

"On the plantations."

"I know what it's like," murmured Harriet.

"It must be horrible."

"Even when it's tol'ble, it's horrible—to someone who ain't free."

Willacassa put her arms around her and wept. "To think I almost married into the system."

"Well, you didn't, honey. That's what counts."

"Isn't there anything I can do, Harriet. Shall I join the abolitionists? What can a white person do? It's not just salving my conscience. I want desperately to do something!"

Harriet pulled away from her, but did not let Willacassa go. Instead she looked up at her, searching hard for the character of Willacassa's concern. And Willacassa was surprised to see so stern a visage. Suddenly the tables were turned. She was the supplicant, the beseecher, and the short, squat black woman before her was the all-powerful, potential bestower of approval.

"You're not playactin' now?" Harriet asked. There was not a soft note in her voice.

"Playacting?"

"I don't think you are. But I've got to be sure."

"Of course not."

"And thisn't a momentary feeling, the settlin's of a disappointed heart?"

"What are you talking about, Harriet?"

"I've got to know." The black woman had not yet relinquished her grip.

"No, it's not a momentary thing," reflected Willacassa. "It's—it's rather like a new religion."

Harriet gently released her hold.

"Then let's talk—in my room."

HANNAH WARREN

Hannah Warren was a spare, long-nosed, almost emaciated widow who lived alone in a house just north of Higbee's Landing Road. Unlike most of the houses in the relatively isolated area, her old Colonial was hidden from the road by a cluster of trees. In fact, her patch of land to the rear and all the land beyond, extending west to Delaware Bay, was a thick forest of pine, oak, and juniper. What farming her late husband used to do was performed across the road on a few acres of flat, sandy soil.

Mrs. Warren did not farm her land anymore except to raise a few vegetables, mainly corn, tomatoes, and green beans, for her own table. Rather she sustained herself by quilting for the hotels and by jarring strawberry preserves and beach-plum jelly.

A tall woman, she wore low shoes with hand-knit stockings showing through below her dress when she appeared at the porch door.

"Yes."

"I'm Willacassa Harrah of Ocean Street on Cape Island. May I have a word with you?"

"About what, my dear?"

"About railroads, you might say."

Hannah Warren's was a hard, penetrating glance, but she said nothing.

"Harriet sent me, Mrs. Warren."

The expression on Mrs. Warren's face thawed and even gave way to a few wrinkles of smile in an otherwise flawless complexion.

"Come in, my child. I was just getting up some tea."

PATRONYM

Willacassa spent a great deal of time at Hannah Warren's house these days. The widow was a solitary woman but full of interesting stories and an excellent hand at tea and cakes. They never talked about the subject closest to their heart—except at the last moment when action was called for. And then they were very business-like, going over the details of their project with thoroughness and precision. But these occasions were rare. For the most part visits by Willacassa were social affairs and time at the widow's house passed pleasantly.

One day when Willacassa was about to leave, she heard someone at the widow's back door.

"Before you go," said Mrs. Warren without rising from her rocking chair, "there's someone I want you to meet. I believe you know each other."

Willacassa turned to see Eliot enter the room.

He had changed very little in the almost two years she had last seen him. He was still a strikingly handsome man with a full head of rich dark hair and finely wrought features. But there was a tentativeness about him, an almost enigmatic reserve.

"Eliot! It's so good to see you again! What a delightful surprise! I had no idea you were on Cape Island!"

Though he was pleased by her reception, it was Hannah Warren who spoke for him.

"You and Eliot appear to have a common interest," she said.

"We do?"

"Yes, the same thing that brings you here brings him to Cape Island. It would be best, of course, not to be too specific. The less we know of each other in this work the better for all concerned."

Ah, the Underground Railroad. Did Eliot know Harriet too? Willacassa's interest was whetted.

"Mr. Irons only recently arrived from New England. When he heard me speak of you, Willacassa, he insisted on renewing old acquaintance."

"Mr. Irons?" Willacassa turned thoughtfully away. "Then your name is Eliot Irons."

"That's right. Anything wrong with that?"

"Irons? That's not a common name, is it? I knew a sea captain by that name once."

"Fairly common in England, I'm told. Not so common in New England."

"Probably a foreshortening of Ironsmith," speculated Hannah

Warren who, it was clear, was fond of the young man in her plain, gaunt way.

"Ironsmith?" repeated Willacassa. "I hadn't thought of that." She wanted to pursue the matter of the name but hesitated to go on. Besides, she was truly happy to see Eliot again, no matter what his patronym.

"Can I accompany you to town?" asked Eliot.

Willacassa was warm to the idea. But Hannah Warren put a damper on things. "It would be best not to be seen leaving together. Why don't you meet each other in town—quite 'by accident'?"

"Perhaps you're right, Hannah. At R. M. Bickcom's Bakery, Eliot?"

"The bakery it is. I'll meet you at one."

When Willacassa "bumped" into Eliot at Bickcom's and they left together, she asked, "How does Cape Island look to you these days? You never did see it in summer."

"Reminds me of home somewhat. Not so much the hotels, which are grand enough though somewhat different from Newport mansions. No, it's the strand and the bathers that remind me of home."

"Newport, Rhode Island?"

"Yes. I'm the son of Captain Irons you met some years ago."

Willacassa surmised that Eliot Irons knew of his father's illegal traffic.

"My father was a slaver, Willacassa. The worst kind. He amassed a fortune running slaves to the South." This admission was difficult for Irons but he went on. "I did not share his attitude towards slavery. Nor do I share his fortune. You do understand that, I hope. In fact, I'm in this work of ours to—"

"Make amends?"

"I can never make amends for what he did. I can only lead my own life."

"I'm sorry. I didn't mean to open up old wounds. It's just that I couldn't reconcile the man your father seemed to be—at least to a child like me—with what he had done."

"What he had done? You mean the slave running business?"

"Yes, that's what I mean," said Willacassa, allowing the matter to rest.

She saw quite a bit of Eliot Irons after that, occasionally at Mrs. Warren's house but more often in town where they would meet "quite by accident"—at R. M. Bickcom's Bakery or at the Post Office on Washington Street or at Canning's News and Express Depot.

"You're partial to Frank Leslie's Illustrated Newspaper?" asked Irons when he saw her buying a copy at the counter.

"I like the illustrations. I sketch some myself and learn a lot from them."

"I see."

"But for reading itself, it is hardly elevating. Though it's no better or worse than Harper's Weekly or Byam's Lady's Newspaper."

"There are some books and pamphlets I have you might be interested in—though not so much for the sketches as the content."

"I've read *Uncle Tom's Cabin*," she teased.

"No, I was referring to Hinton Helper's *The Impending Crisis of the South*, Frederick Douglass's *My Bondage and My Freedom*, and some speeches by Wendell Phillips."

"You don't have to make a convert of me, Eliot. The fact that we both meet at Hannah Warren's house should be proof enough of that."

"I'm sorry. That was not my intention."

"And if you are trying to educate me, you should know that I received a fine education at the Cold Spring Academy. Reverend Williamson can attest to that."

"I'm afraid I've been somewhat insensitive."

"I'm afraid you have."

"Then let me make amends, after all."

"Amends?"

"Let me show you that I'm not all bad—even if I am an Irons."

"And how do you plan to do that?" asked Willacassa.

"I'll read your fortune."

"My fortune?"

"Well, your palm. And if it's as lovely as your face, it should tell a fair story."

Willacassa found this light side of Eliot Irons much more to her liking than the proselyte in him. But she was not about to turn her hand over to him.

"Let me think about it," she said. "After all, I still hardly know you. It won't do at all to let you read my palm before I read *The Impending—*"

"*Crisis of the South?*"

"Yes, that's the one. And then we should more or less be on an equal footing."

"Whatever you say, Miss Harrah."

But Eliot Irons never got to read Willacassa's palm. On Saturday when she was supposed to meet him, Willacassa got word from Hannah Warren that she was to be stationmaster that day.

REVELATION

Later in the week they did get together, meeting at Canning's, and Eliot proved to be a good companion. He had the gift of lifting Cape Island out of its provincialism into the ever-widening circle of the country at large. What was happening in Washington, Philadelphia, New York or Boston suddenly became of paramount interest.

When some lurid accounts of Mormon polygamy appeared in a Philadelphia newspaper, Irons protested against the sensationalism.

"I'm not one for polygamy. But I seriously doubt that the Mormons are as lewd and licentious a group as made out."

"I really don't know," said Willacassa.

"After all, polygamy was not unheard of in Biblical times."

"That's true. But—"

"But what?"

"I should think it would present problems—domestic problems."

"Better let the Mormons worry about that," declared Irons. "I can think of a few domestic problems in families where monogamy holds reign."

"In any case, it was not an issue to wage war over," pursued Willacassa.

"The Utah expedition was hardly a war," replied Irons.

"Yes, I read the comment, 'Wounded, none; killed, none; fooled, everybody.' But the military action resulted in much dislocation and a good deal of harassment."

"I suppose it did," he said, conceding new respect for Willacassa who in some way he regarded as his protégée.

For her part, Willacassa was coming to a new realization. Her contact with Eliot had opened fresh vistas for her. His concerns had become her concerns. There was a broadening of interests that meant there could be no turning back. The days of sand and surf had given way to more earthy problems. She could no longer respect a man who was not her intellectual superior.

Willacassa recognized that this expectation was neither fair nor prudent. For most of the men she met simply did not measure up to her standard. Either they were masters of what were presumed to be male prerogatives: horsemanship, marksmanship, or gambling (and the concomitants of gambling—drinking, smoking, and bluffing). Or they were veritable simpletons who tried to clothe with fashionable dress and cliché speech their intellectual nakedness.

In Eliot Irons she saw what until now only Reverend Williamson and her father represented—intellectual curiosity. But of course
they were of a different generation. In her own generation only
Eliot Irons seemed worthy of interest. Only Eliot Irons carried the
torch of contemporary awareness. The others could not even ig-
nite their intellectual sulphur matches.

And so she prepared for Eliot to make his move. There was
every indication that he was taken with her. Even allowing for a
natural shyness, a reticence on his part, she sensed that he was
hopelessly attracted to her. Though he was at times hesistant when
he was close by, even unsure of himself, she had no doubt that
he would overcome these obstacles and begin his courtship. But
Eliot was so unpredictable even she dared not hazard a guess as
to what form this courtship would take.

One evening when they had attended Independence Day
festivities on the lawn of Congress Hall, Eliot drew Willacassa away
from the crowd watching the fireworks and led her in a direction
away from the ocean. He strolled quietly, making no statement or
gesture except to offer his arm, which she took.

After a while they found a small gazebo outside one of the
newer establishments and sat down.

"Do you recall that I once said I had made a mistake?"

Yes, Willacassa did recall. But she did not feel it was her place
to recollect the matter.

"Well, I have since taken measures to rectify that mistake."

Eliot looked as though his life depended on what he was going to say next.

"You see, Willacassa, I was a married man when I stayed at
the Mansion House."

"Married?" This indeed was more than Willacassa had bar-
gained for. She did not know how to react to the news.

"But the marriage has since been annulled."

Willacassa struggled for something to say. "You—You don't
have to account to me, Eliot."

"But I want you to know how it happened. A terrible mistake
was made."

"We all make mistakes, Eliot."

"Yes, we do. And the fault was as much mine as hers. You
see, we were both part of the movement."

"The abolitionist movement?"

He nodded. "Yes, but that's only one label. Our chief purpose
was providing an escape route for runaway slaves. And among the
circle of people I met with regularly, Emmaline was by far the most

brilliant. It did not matter that she was fifteen years my senior. I was seduced by her intellect.''

He paused as a strangely perplexed look crossed Willacassa's face. ''But intellect makes a poor substitute for a bed companion,'' he said at last. ''It wasn't that she was older than I am. Some older women are deucedly attractive. I'm afraid she wasn't blessed at any age—except with that keen intellect. And it is difficult to go to bed with strings of words, no matter how many pearls of wisdom are attached to them. Abolition was the matchmaker, but in every other way we were misfits. I deserted her that first night.''

''You needn't go on. Eliot.''

''I'm telling you all this so that you will understand why I did not express my true feelings to you two years ago—before I left so suddenly.''

''But I thought it was the bank failure drove you away!''

''It was. I had been entrusted with a lot of money to help the Underground Railroad in South Jersey. And all at once it was beyond reach. But the bank failure alone did not prevent me from returning last year. By then a good part of the money had been restored by the state. It was the marriage that kept me away. Until it was annulled, I felt I had no right to press my case.''

Willacassa shook her head in bewilderment. ''Oh, Eliot! You confuse me so! I don't think I'll ever understand you.''

''But—''

''No more! No more! At least not tonight.''

With that she hurriedly stepped down from the gazebo and started home alone.

RESENTMENT

Eliot's revelation threw Willacassa into utter confusion. Without admitting this to herself, she had hoped that Eliot's presence on Cape Island would fill the void left by Francis Grandee. A terrible void that nothing seemed able to fill, not even her work with Hannah Warren.

In fact, since encountering Eliot again, Willacassa wondered whether there would have been a romance with Francis had Eliot remained on Cape Island. She surprised herself with speculation of this sort, given the depth of her attachment to Francis. But there was no denying Eliot had made a strong impression on Willicassa's young mind when he first appeared as caretaker of the Mansion House. And his mysterious comings and goings only added to that impression. More important, he had come on the scene during the off-season. This boded well for a lasting relationship. No ten-week romance here. No farewell at the Landing in September.

But she resented that there was someone in Eliot's life before her, even Emmaline—if that was the unfortunate lady's name. At the same time, she realized that resentment of this kind was irrational. She preferred to think of herself as a reasonable person, held in high regard by people like Reverend Williamson and her father. But try as she would she could not drive this resentment out of her mind.

And so she did her best to avoid Eliot. Irons, of course, would be anxious to see her again, to set things straight. He had much to overcome, and it was natural that he would try to make Willacassa see things as he saw them. But she desperately needed breathing room. She needed time for things to settle in her mind. She tried to avoid the shops he was likely to patronize. Despite her efforts, he ran into her at Canning's where she had gone to buy some newspapers for her father.

He managed to accompany her outside. "Did what I tell you the other night put a barrier between us, Willacassa?"

"I'm sorry, Eliot. But I still haven't gotten used to the idea."

"You think less of me for having made a fool of myself?"

She tried to respond but, finding herself unable to, simply nodded.

"Well, I can't say you're not forthright," observed Irons.

"I'm sorry, Eliot. It's my fault. But I can't understand how a man of your intellect could do such a thing!"

"But that's the point. It was intellect that was at fault. I should have followed my instincts. Though it was a bit late in the day,

I finally did follow them.''

"When you left her?''

"Yes.''

"As you say, it was a bit late in the day.''

Seeing the pain in his face, she checked herself, then reached out. "Oh, forgive me, Eliot! I'm being so unkind! You don't deserve such criticism from a friend!''

He took her hand. "But I do. And you haven't been any harder on me than I have been on myself. It was a stupid thing I did. And in the end a cruel thing. But there seemed no other course.''

The word "friend," used by WIllacassa, had not gone by unnoticed. Irons had hoped he was more than a friend to her, but under the circumstances settled for what he could get.

"Then we can put the matter behind us? I know I have a way to go before I—''

"There's no point talking about that now. But we are friends and other concerns will have to occupy us. Hannah Warren wants to see you tonight. She was unable to contact you at your old lodging.''

"I had to move. Some new arrivals rented the house for the month of July. My room was needed too.''

"Where are you staying now?''

"At Hotel Purgatory—until I'm readmitted to Paradise.''

Willacassa's was a reluctant smile.

"I'm afraid Hotel Paradise is booked solid for the season. But even if it weren't, I'm sure it would fall short of your expectations.''

A week later it was Willacassa's turn to call at Hannah Warren's house.

The widow met her at the door. "Eliot's here. He's been here the last few days.''

"Is anything wrong?''

"He's been hit by a ball and refuses to see a doctor.''

"What happened?''

"I don't know the details. Except that he was shot. It doesn't seem to be serious, but any wound needs treatment.''

Eliot was sitting up in a chair, reading a newspaper, when she entered the small room upstairs. His arm was wrapped in bandages.

"Not the same arm, Eliot?''

He put the newspaper down. "The same as was burned. If you're here to dress it, Willacassa, you're too late. Hannah beat you by three days.''

"You should see a doctor, Eliot.''

"And give the whole scheme away? Come now!''

"Was the trip completed?''

"Exactly as planned."

"You're a brave man, Eliot."

"No braver than you or Hannah or Harriet. In fact, it's Harriet that deserves the credit. How that woman manages to slip in and out, I'll never know. And how she persuades the Blacks to seek their freedom defies belief. But the less said the better."

They were joined by Hannah Warren.

"The operation is becoming too risky," she said. "Government agents are all over the county rounding up runaways. They are more determined than ever to shut us down."

"Even shoot us down," added Willacassa.

"They can't catch all the runaways," affirmed Irons. "And there are more coming every month. What was once a trickle is now a stream."

"All the same, we must be careful," the widow cautioned.

When Hannah returned downstairs for a pot of tea, Willacassa took the opportunity to deal with Eliot's convalescence.

"If you'd feel safer, we could put you up at Trescott House and I'd stay here. I could tell father you were in an accident."

"Thank you. But a guest house—with all those people about—might not be the ideal hiding place. Besides, once I remove the bandages, nobody will be the wiser."

Willacassa put her hand gently on his other arm. "Eliot, I've been a fool. Forgive me!"

"For what?"

"For being so narrow-minded. For having so little tolerance for what is after all no more than human error. I'll try to be more enlightened in the future."

"There's really nothing to forgive." he said quietly. "Though it's good of you to think so."

GOOD COMPANIONS

Before too long they were friends again. They shared views on Hinton Helper's *The Impending Crisis of the South* and on the speeches of Wendell Phillips.

"Why don't *you* try your hand as an orator?" suggested Willacassa.

"I've tried the lyceum platform, but I'm no Wendell Phillips."

"You've heard him speak?"

"Several times. He's a man of great charm and persuasive power. And he's an impassioned speaker, which I am not. Always I saw his wife at his side. A spendid partnership, if ever there was one. People say it wasn't until they married that Phillips caught fire."

Irons seemed to come alive when he talked about Mrs. Phillips, the former Ann Terry Greene. "Even when his family regarded him as a fanatic, she was always supportive of Phillips. She was a pretty woman, an ideal complement to his patrician good looks. And the one who urged him not to affiliate with any political party."

It was exciting to see Eliot in such good form. But Willacassa found all this enthusiasm a little troublesome, though she could not say why.

"So I put what talent I have into the Underground Railroad," concluded Eliot, "and left the lecture circuit to others."

This remark was so typical of Eliot, so self-effacing that Willacassa was almost annoyed by it.

"Come now, Eliot! You do yourself an injustice. Eliot Irons is a fine man in his own right. He needn't be copy in silhouette of Wendell Phillips."

"Well, we do disagree on one thing."

"And what is that? The vote for women?"

He shook his head. "No, prohibition. I don't think it can be made to work."

"Why not?"

"I haven't figured out the reasons yet. It just seems to go against the nature of man."

"Of some men," she corrected him.

"All right, of some men. Besides, the whiskey tax keeps this country solvent. If there is no whiskey, there is no tax."

"I don't know whether to take you seriously or not."

Irons was the picture of sobriety as he looked at her. "You must take me seriously, Willacassa. You must always take me seriously."

At the summer's end they picnicked near the water. Towards the east the beach was all but deserted except for the dunes and the everpresent screeching terns, although to the west the hotels were still doing a good business.

Eliot sliced the sourdough bread while Willacassa pulled out the cheese and the wine from her basket.

"It's been a good summer." announced Willacassa. "Father said he made more money this year than any time before. There was not a room to be had any week this season. There is a problem though."

"And what's that?"

"The mail. The stagecoach comes three times a week but service is irregular. The *Ocean Wave* is urging daily mail."

"Makes no difference to me," said Irons. "I receive letters from no one. Not even you."

"Me? But there's no need for me to write!"

"Oh, yes, there is! You never tell me that you care for me. If it's shyness, you could always put pen to paper."

"Well, it isn't shyness."

"Then what is it?" he asked, following a curlew circling in the sky.

What was it? She did not know herself. She knew only that what troubled her was a general feeling about men, not Eliot in particular. It was a suspicion that men were weak, or at least not as strong as she had been led to believe. They did not, when it came to standing up at the critical moment, show strength—strength of purpose or strength of character. To use a phrase of her father's, they showed "No starch."

Oh, there were exceptions. Her father for one, although she could not abide his everlasting neutrality with respect to his guests. And Reverend Williamson. And Joseph Leach, editor of the *Ocean Wave.* But these men were not her contemporaries. Maybe they were a different breed a generation ago! Maybe someone had thrown away the mold! Except for Henry Sawyer, who charged at life the way a hussar charged a barricade, all the men she knew had flaws. (Even Eliot who was brave enough and certainly exhibited strength of purpose.) Not tragic flaws in the Greek sense. Just flaws.

She handed Eliot a cut of cheese and the bottle of wine.

"See if you can open it," she said in no hurry to answer his question.

Eliot applied a corkscrew and slowly pulled the cork free.

"I'd ask if there was someone else," pursued Eliot. "But I pretty much take up all of your time."

"That's right, Eliot. I have little time for anyone else. And what time I have must be spent at Trescott House."

"Unless—"

"Unless what?" By now she had put some cheese and bread together for her own plate.

"Unless you're conducting a secret romance with one of your father's guests."

She laughed. "Father's guests? All they do is smoke, drink whiskey, and play cards. The nearest thing to a ladyfriend is the queen of hearts."

"Then I have a chance? You'll not hold Emmaline against me?"

A certain mischief crossed Willacassa's face as she replied.

"Since you did not hold Emmaline against you, I won't either."

1859

A LOVER RETURNS

TRANSFORMATION

Willacassa had been looking to escape the heat and had stepped into People's Ice Cream Saloon on Washington Street, between the United States Hotel and the Washington House, when she saw him. After an absence of two summers, Francis Grandee was once more a visitor on Cape Island. Her heart turned to butter. She swung around to avoid being seen by him, but too late. He tapped on the saloon window and, when she did not respond, entered the bell-ringing door.

"May I sit down?" he asked, his well-molded face no less handsome for a lack of sun.

"I'd rather you didn't."

"But I must—I must speak to you."

"Then sit down. Only let's not have a scene."

"How are you, Willacassa?"

"Quite well, thank you."

"Did you miss me at all?"

"Miss you! I don't think about you at all anymore."

Even sitting down, Francis Grandee seemed shorter than she had imagined. But there was no loss of confidence in his ability to charm people.

"I've changed, Willacassa. A marked change has come over me."

"You don't rape pregnant women anymore?"

"That wasn't kind."

"Nor was it meant to be."

"Well, the truth is, I don't. In fact, I've had a change of heart about slavery. I don't advocate it anymore. I see it as an evil, a necessary evil, but an evil nonetheless."

Willacassa did not interrupt. She was anxious to see how far Francis would go with his "change of heart."

"You see, I've given it a lot of thought. And I've spoken to a lot of people. There are people in the South who oppose slavery, you know. And the more I've thought about it, the more I realized how restrictive it is to the economy of the South. But, of course, there is no moral justification for it either. I'm aware of that now. These poor blacks should be left alone. And, if they wish, shipped back to Africa."

"I don't believe you, Francis."

"You don't believe what?"

"That this is a sincere change on your part. It all smacks of Southern arrogance."

"But I am changed. I don't countenance slavery anymore. Is it arrogance to suggest that it isn't the overriding issue between North and South, at least for me? What I want is States' Rights, freedom from the North. I would free the slaves tomorrow if I could. Just as I want the South free of Northern restraint."

Willacassa evinced a faint sneer.

"It doesn't sound all that different to me. States' Rights is just another way of doing what you want."

She accepted her ice cream from the waitress and began eating it while Francis ordered a plate for himself.

"Chocolate, please. No vanilla. I never mix the two." Even he had to smile when it dawned on him that a new understanding could be put on his intended meaning.

"You see, Willacassa, I'm a new person. Not the man you were engaged to, but a totally new being. So much so that I've had a falling out with my father."

"He still champions slavery?" asked Willacassa.

Francis Grandee nodded confirmation.

"And he's all but disowned you?" she mocked.

"Not exactly. But he's not at all happy with my decision."

"Decision?"

"Decision to run for the South Carolina legislature as an opponent of the status quo."

"You mean slavery?"

"Well, the status quo includes slavery, doesn't it?"

"I suppose it does." She thoughtfully took another spoonful of ice cream, unable to make up her mind about what Francis was saying.

It was evident from all he said and the way he presented himself that some kind of transformation had taken place in the man. But Willacassa saw no superficial difference in him. The cast of his eyes, the short nose, and the strong but slightly receding chin were all the same as before. Nor were there lines of character in his face to deepen or shade the impression of youthfulness that he habitually conveyed. Willacassa expected something more from Francis. Transformations should show in one's expression. But all was the same with him. And so his motivation and his sincerity were suspect.

Francis could see he was making no headway with her. "How can I convince you that I've changed?"

"You need not trouble yourself. If you have changed, then you should be the better for it. But it makes little difference to me. I have not forgotten what took place that night. Nor am I likely to forget it. Any feeling I had for you once is gone."

"Is it?" he asked huskily.

She nodded emphatically.

"And it can't be rekindled?"

"No, it cannot. You see, I've changed too. We're not a likely couple anymore, Francis. Nor will we ever be,"

"Willacassa, you can't mean that. There must be some residue of feeling left for me."

"If I told you that," she said. "I'd be lying."

"Can't we see each other at all?"

"There are other girls on Cape Island to see. Some—especially those from the South—might be interested in your transformation. I'm not. Besides, I'm busy these days. Summer may be holiday time for visitors, but for the year-rounders it's nothing but work."

She rose to go.

"Mother sends her love," said Francis as she moved away from the table.

"Thank you," said Willacassa, and with that she pulled open the door, ringing the bell connected to it.

A STUDY IN CONTRASTS

Willacassa was greatly agitated by Francis Grandee's appearance on Cape Island. Without admitting it to herself, she had looked every day for his return the summer of their estrangement and, when there was no sign of him or of any of the other Grandees, she had looked again the following year. Not that she would have forgiven him for what he had done, or could forgive him. It was simply that she could not believe she meant so little to Francis Grandee as to be dismissed from his life without a struggle.

When in his absence Willacassa had finally reconciled herself to the idea that he was gone and that she was truly the better for it, Francis Grandee perversely popped up. It was upsetting to her, his return. Willacassa could not deny there was an emotional residue from the affair. It clung to her like powdery, dried-out honey, and she could not scrape if off. Nor was she entirely sure she wanted to.

Then there was Eliot Irons. Though she had her doubts about him since learning of his aborted marriage, she had come to see him in a new light. While others, like the much-admired Wendell Phillips, talked—talked with passionate eloquence, she admitted—Eliot acted. (She assumed Hannah Warren gave him the more dangerous assignments.) They were, moreover, of the same mind on many things. John Brown's raids into Kansas and Missouri, though fraught with violence, seemed to them to be the only way that slaves would eventually be freed. Susan B. Anthony's Women's Rights Convention in New York City was the beginning of a much-needed feminist movement that would ultimately free the nation of its white slaves. And the discovery of gold in Kansas and in the Washoe Mountains of the Utah Territory would only rid Cape May County of its drifters, adventurers, and ne'er do wells, much as the California Gold Rush had done.

And, of course, there was a tacit understanding that the work they were doing—each in his own way—was of such overwhelming importance that nothing should stand in its way. But it did not take a clairvoyant to see that the mere presence of Francis Grandee on Cape Island could be a threat and a danger to them.

For Francis Grandee was Dixie as much as Dan Emmett's widely popular new song was "Dixieland," and if corn was more the crop at the Grandee Plantation than cotton, it made little difference. His great transformation notwithstanding, Francis Grandee would take his stand to live and die in Dixie. And in no way was Willacassa going to allow the work of the past two years to suffer.

So when he sent her a box of Whitman's candy or a dozen yellow roses or a bottle of imported cordial—with a note unsigned—she did not so much as sample the chocolates or sniff the roses. And when she turned the bottle of cordial over to Rachel, the good woman took it without hesitation.

"I saw Francis Grandee today," noted her father. "I suppose you've seen him too."

"How did you know?"

"I assumed he wasted no time making himself known to you."

"He met me at People's."

"And what did he have to say?"

"Nothing of any consequence."

"Do you still care for him?"

No one could say her father wasn't direct.

"I think not. But I'm not sure."

"An honest answer."

"Would I lie to you, Daddy?"

"No." said Harrah.

Eliot Irons called. He had picked up the issue of *The Atlantic Monthly* he had meant to show her. In particular, an essay by Oliver Wendell Holmes who had recently published "The Autocrat of the Breakfast-Table." But there was also an article on the Senator from New York, William Seward, that he wanted her to see.

"They say his uncompromising stand on slavery will cost him the Republican Party nomination for President in 1860."

"How uncompromising is it?"

"He insists the United States will sooner or later become one or the other, a slave-holding nation or entirely free labor."

"The politicians won't buy that," declared Willacassa.

"Of course not."

"They'll find someone more diplomatic, less controversial. No one who meets the question head on will be the standard-bearer."

"That's a certainty," said Irons.

Leaning over the counter behind which Willacassa drew up menus and bills, he tried to whisper in her ear. Thinking he wanted to kiss her, Willacassa drew back.

"You needn't worry," said Irons. "I wasn't going to kiss you. I only kiss women who are passionate about me."

"I do like you, Eliot." She kissed him on the cheek. "In fact, I like you more than I care to admit."

This was quite true, if quite different from her feelings of the year before. Apart from her initial response to his good looks, a new attraction had developed. There seemed after all a certain inevitability about Eliot's being on Cape Island, and inevitability was something Willacassa had difficulty resisting. It appeared altogether fitting that the son of Captain Irons would gravitate to Cape May County and be a moving force in the Underground Railroad. It made up for a terrible wrong. But the other aspect to this was Captain Irons himself.

For Willacassa had always had ambiguous feelings where the master of the *Carolina* was concerned. Though she had seen the bloated bodies of the slaves on Higbee's Beach and knew that this was the captain's handiwork, she could not reconcile the genteel, soft-spoken New Englander with what he had done. He had talked like a civilized man. He seemed fatherly and caring. It was not possible he could be guilty of such an outrage. And yet all signs pointed to his guilt; of that there was no question. Willacassa had trouble dealing with this as a child. And she still had trouble dealing with it. For in his way Captain Irons had sunk his hook into her child's psyche. He had called her his "Angel of Cape Island," a most flattering appellation. He had promised to return and indeed did return to Cape Island, laden with gifts. But in the interim, of course, there had been that terrible discovery on Higbee's Beach. And Willacassa had recoiled in horror. She did not know what to make of such sinister contradictions. She did not know how to cope with her feelings. It had been most difficult for the girl to turn her liking for Captain Irons, at least during his short stay at Trescott House, into dislike. But with the arrival of Eliot, whom she had been attracted to in his own right, she eventually found a way to release the troubling affection she had so long suppressed. By shifting her affection to Eliot, Willacassa was able to divide her feelings and sort them out. By transferring the best of Captain Irons to his son, she could finally take the captain to task.

But Eliot proved a problem too. As decent a man as he was, she was not sure he could sustain the burden of affection transferred. He was after all only the son of Captain Irons, not Captain Irons himself. To pretend otherwise was a pretty piece of self-deception. As a suitor, he had to sink or swim in the sea of his own persuasion and not cling to what was left in the wake of the *Carolina's* wreckage. Willacassa knew all this. She was not a fool. But she too had a difficulty swimming in the rough waters of revealed truth.

And now the reappearance of Francis Grandee to make matters worse!

"What I really wanted to say," said Irons whom she realized was still close to her ear, "was that despite the heat, you look very fetching. Would you care to have ice cream at *People's* ?"

Willacassa's was a half smile as she put to rest all the thoughts that had been whirling about in her head.

"Oh, Eliot! I've already had ice cream, thank you. More than I can handle for one day."

OUT OF THE FOG

She awoke to a smoldering fog which lingered into midday. Eliot appeared at Trescott House and suggested they visit the archery and pistol tents as no one was on the beach. She was not particularly inclined to mix with the meandering crowds, but she did not want to disappoint Eliot.

As Willacassa feared, the crowds were enormous. But few people actually ventured into the tents, preferring the novelty of the thick fog. As Eliot stepped inside to pick up a pistol, Willacassa thought she saw someone lurking nearby. But she concluded it was only the mist playing tricks on her.

Eliot took aim and fired at a small target, cutting a hole very near the bull's eye.

"That's good shooting, Eliot. My brother Jonathan would have been proud of you."

"It is an improvement over last week's performance. The way things are going, I may carry a pistol next trip out."

"Things getting that bad?"

"Yes, you'd think they were giving bounties the way these hired men pursue their quarries. Particularly up Salem way." He paused suddenly. "But the less said the better."

They turned to see a young officer shoot some holes through a bull's eye as he rapidly changed empty pistols for the proprietor's loaded ones.

"If a war breaks out," observed Eliot, "he'll be ready."

"Yes," concurred Willacassa, waving the acrid smoke away, as the officer's companion drawled approval. "But on which side?"

As they stepped out of the tent and headed for the lawn of Congress Hall where a refreshment stand had been set up, they bumped into Francis Grandee.

"Willacassa," nodded Francis, his glances almost furtive, "may I have a word with you?"

Eliot was taken by surprise. But something—Willacassa's sudden coloring perhaps—assured him that the stranger had been an important person in Willacassa's life. Gallantly stepping aside, he tipped his hat and said something about having to buy some newspapers.

"Why didn't you introduce me to the gentleman?" asked Francis when Eliot had disappeared into the crowd.

Willacassa convinced herself that she had avoided the social niceties of the situation to lessen the danger of exposure for Eliot. But it was possible she had other reasons which she did not want to dwell on just now.

"Oh, he's not that close a friend," she pretended.

"I see. But not for lack of trying, I would hazard."

Willacassa decided it would be best to ignore the remark, though she gave him credit for being perceptive.

"I know you don't believe me," Francis began. "So I'd like to tell you how my transformation came about."

"I'm not interested in your transformation."

"But it's important you understand."

"No, it's important that *you* understand. What you do or have done no longer concerns me. I want only to be left alone."

She found it disconcerting that though the fog still shrouded most everyone else from view, Francis on the lawn of the Congress Hotel stood out plain and clear. It was as if he had somehow persuaded the elements to aid him in his endeavor which, of course, was to make it appear that nothing else mattered, only this business of his transformation.

"Won't you at least hear me out?"

Willacasssa was tempted to let him tell his story if only to be able to dismiss it out of hand and be rid of him. But she toughened her stance and simply said, "You're wasting your time, Francis. Nothing you have to say will make one whit of difference!"

And with that she too melted into the crowd.

"What did your friend want?" asked Eliot when Willacassa found him at a book stall which was raising money for county orphans.

"To resurrect the dead," replied Willacassa.

"What did you tell him?"

"That it was impossible."

"I thought nothing was impossible—for you, Willacassa."

"You thought wrong, Eliot."

GRANDEE'S LETTER

Francis Grandee now began to stalk her in earnest. Since meeting Willacassa on the lawn of Congress Hall, he had managed to cross her path on a few occasions, not a difficult thing for him to do as he was staying at Columbia House just across the street. But as if by prearrangement he met Willacassa coming out of Jesse E. Smith's Clothing Store, diagonally opposite Congress Hall, where she had gone to buy a cap for her father's birthday. The next day he found her in front of S. R. Ludlam's Confectionary Store. The day after that he was seen at the ice cream saloon at the Steamboat Landing, opposite the Oscar House where Willacassa and her father had gone to pick up a birthday cake for Rachel's birthday, these widely different personalities, her father and Rachel, having been born within the same week in August. But on none of these occasions did Willacassa give Francis Grandee so much as a glance.

One day Willacassa called at the post office when she saw her name in the *Ocean Wave's* "List of Letters," letters remaining for the week of August 12, which the post office wanted picked up.

"An advertised letter, please."

"Oh, yes, Miss Harrah," said the clerk. "Here it is. A gentleman left it."

"Left it?"

"Yes, he paid the postage and asked that it be listed."

"Isn't that a bit irregular?"

"Well, it isn't regular. But it is done."

Of course it was from Francis. She stepped out of the post office and ripped open the letter.

"You needn't bother reading it," said Francis Grandee, emerging from beside the door. "It says I still love you. But it was no more than a pretense for me to meet you."

"You've been waiting all this time?"

"Since the office opened at 6 A.M."

"And what if I didn't come until 8 P.M., closing time?"

"I would have been older and wiser, but still would have waited."

Willacassa was pleased by what he said, but she persisted in her interrogation.

"Don't you know that I have other, more important things to do? I thought the letter might be important."

"It is," said Francis Grandee, starting to pursue her as she quickly made her way across the dusty street. "May I give you a lift? I have a carriage waiting. I can tell you all about the letter as we ride."

"I much prefer walking," said Willacassa. "That's the beauty of Cape Island. A person can walk from one end to the other without so much as needing a conveyance."

"But look how dusty you'll get."

"The price of independence. And it's well worth it."

When she returned home, Willacassa regretted not having taken his offer. She was footsore and her gown was covered with dry August dust.

"Eliot was here looking for you," said her father.

"He's back?"

"I didn't know he was away."

"Yes, he had to leave for a few days. I think he went to Dennisville to look at some boats."

"Dennisville? Aren't there boats enough at Cape Island?"

"He wants to see how they're built. He may go into boat-building himself one day."

Harrah smiled at his daughter, then put on his new cap to go out.

"How does it look?" he asked.

"Handsome. Makes you look ten years younger."

Harrah peered into the full-length vestibule mirror.

"Ten years younger, eh? Not bad for a man of fifty-three." Not bad at all, thought Willacassa. Tall, trim as a schooner, with barely a trace of gray at the temples. She wondered whether Francis Grandee would look half as good at his age.

When she was alone in her room, Willacassa read the letter she had stuffed in her handbag. She might not know the transformed Francis. But she knew the old Francis well enough to assume that, though he dismissed it as a pretense for meeting her, he regarded it as a significant utterance and wanted very much for her to read it.

"Dear Willacassa,

It is apparent by now that you do not believe in my transformation. This troubles me greatly. If my new life means nothing to you, then how can I embrace it as fully and devotedly as I should like?

You think the transformation was a hasty affair, an empty and hollow gesture, worthy of the head-strong son of a Deep South plantation owner. If you think that, I really cannot blame you. Left to my own devices, I would have deserved your contempt.

But something happened to me that changed things, that ended past practices forever. I hesitate even now to tell you about it. But as you seem to doubt me, I am compelled to tell you the whole story. I pray you will think none the less of me for it.

You may or may not remember the name Noah. Well, Noah was a field hand, a big, hulking brute of a man who could break the bolster of a cart in two with his bare hands. It was Noah who fathered Melva's child and became her man. And it was Noah who collared me on the road one night and threatened to slit my throat for what I did to Melva.

'I could kill ya, Mass Francis,' he bellowed. 'I could kill ya in the wink of a eye. But I won't—if ya makes a promise.'

'Anything,' I pleaded. Yes, Willacassa, I was reduced to pleading for my life and that to a sweating, stinking black man. 'What do you want me to promise?'

'To free Melva and me and the chile.'

'But you belong to my father!' Even in desperation I had to tell the truth.

'Makes no diff'rence. Ya can buy our freedom.'

'I have no money of my own.'

'Then get some! I don't care how.' He seized my throat as if to punctuate his remarks.

'All right,' I gasped. 'I promise.'

But Noah wasn't about to release me.

'How do I know ya'll keep yo' promise?'

'I'm a Christian,' I replied.

'A Christian don't hold slaves or rape women.'

'I swear by all that's holy!'

Noah looked at me long and hard.

'Listen carefully, white boy.' Not once did he relinquish his grip.

'I'm listening.'

'You tain't a Christian now. But yo goin' to be one. Do you hear Noah?'

I nodded, unable to talk by now.

'Ya goin' to read the Good Book. Backwards and fo'wards. Do ya hear? And then you be a Christian. And then you free Melva and the chile and Noah. Is that plain? I'll give ya time, but not too much time. And if'n ya tell anyone about this, I have friends who'll cut your throat same as me. Do ya understan'?'

'Yes,' I gasped.

'And ya'll keep yo' promise?'

'I'll keep it,' I said.

And Willacassa, I did keep it. It took some time, almost

a year. But I finally found someone to borrow money from. This friend of mine acted as an intermediary and arranged for the purchase of Melva, Noah, and their child.

My father balked at first, and I think he had his suspicions. But in the end the money proved too much for him. It was a formidable sum for three slaves, and he could not resist. Later I paid the money back by selling off some thoroughbreds my father had given me. So in a way he paid for them himself.

You may wonder why Francis Grandee kept his word. After all, it was given under duress. The truth is I was frightened, not so much by Noah's strength as by something in Noah himself. If I tell you there was something of the Lord in him, something terrible and awesome, would you believe it? The words literally rumbled out of his mouth, shaking me and the ground at the same time. And when he taunted me about not being a Christian, I felt I was getting the Word straight from the Almighty himself.

Thus my transformation, Willacassa. I am indeed a Christian now. I may never be able to shake my Southern heritage completely. But I will never again hold a man in bondage or take advantage of a woman's defenselessness.

I need only one thing to complete my conversion, however. And that is your forgiveness. Without that, a stain remains on my character which can never be rubbed out. You needn't say you love me—though I will always love you— but tell me that I'm forgiven and I will find the peace I am looking for.

Yours always,
Francis''

OPEN TO CRITICISM

Willacassa did not know what to make of Francis Grandee's letter. Though she had doubts about what she was doing, she turned it over to her father for his reaction.

Harrah read with unflagging attention what young Grandee had written, not once looking up at Willacassa to share his feelings. When he was through, he put the letter down and removed his glasses.

"Well, Father, what do you think?"

"I think we're a generation riddled with guilt. Of course, Francis has more reasons than most for feeling that way. But his is not an isolated instance."

"You mean Eliot?"

Willacassa had never discussed with her father Eliot Irons's reasons for coming to Cape Island. But she suspected that her father had some idea even though he never openly alluded to Captain Irons and his ill-fated ship, *Carolina.*

"Eliot, Joseph Leach, Reverend Williamson, myself—any man who pretends to have a moral standard for himself, if not for others. This problem with slavery—it is such a moral abyss that I'm afraid the whole country will be dragged down before it is solved."

"And Francis?"

"Francis is a Southerner. It goes worse with them. It's bad enough being a poor Southern farmer with no slaves. Or a mechanic with no prospect of work because of slave labor. They support slavery too. It wouldn't be loyal not to. But Francis is part of the slavocracy. To be a Southern plantation owner and not have any moral misgivings about what you're doing is one thing. To be one of the select few who benefit from slavery and suddenly have serious doubts about its propriety—that's pretty heavy water. It's enough to tear a man apart. Or bring about the kind of transformation Francis writes of."

"What about Noah?" asked Willacassa.

"It doesn't matter how the transformation came about. Whether it's out of fear. Or divine providence. Or even a desire to convince you he's now worthy of you, Willacassa. The important thing is that Francis feels he's been transformed. And so important to him is it that you appreciate this, that he is willing to expose his own faults and weaknesses in the process."

"Then I should forgive him?"

"That's up to you, my girl. At least up to a point. I don't see that it's in your power to forgive him for what he did to that slave girl. She's the only one who can forgive him for that."

"But Noah seems prepared to overlook that."

"Noah got a trade off. As far as he was concerned, the deed was done. But at least he got something in return. He got the girl, his child, and his freedom. Not a bad day's work for a man with limited bargaining power."

Armed with her father's logic, Willacassa sat down to write Francis Grandee an answer to his letter. She was not totally satisfied with her father's interpretation of its significance. But she had no better explanation of her own. So she penned what she considered a thoughtful reply.

"Dear Francis,

As far as I can forgive, you are forgiven. But it is not for me, it is for others, the victims of your injustice, to tender forgiveness.

I am sorry to say, however, that things can never be as before. The love I had for you is no longer there. And it cannot be rekindled. I hope you understand that I am not trying to make you unhappy or to crush fond memories. It is just that I must be honest—with you, with myself. You will, I am sure, find someone else to share a new life with.

Willacassa"

No sooner had she posted the letter than she was filled with regrets. Willacassa had not wanted to close the door on Francis. While she did not wish to encourage him, she did not want totally to reject him. She found herself blaming her father for her dissatisfaction. He had been too philosophical about Francis, not nearly personal enough. And he had not told her what she wanted to hear.

"It is easy for you, Father, to talk about Francis and the others. But what about yourself? You wine and dine these Southerners. And a fair share of Northerners, too, who share Southern prejudices. I don't see you suffering any pangs of conscience for that."

"I'm in the hostelry business. I can't discriminate against anyone I serve."

"Why not?"

"It's just not done. These people are guests in my house. I cannot be rude to them."

"No matter what they say or do at home?"

"As long as they don't do it here."

"In other words, you're neutral," declared Willacassa.

"Not really."

"In your behavior towards them, you show the same courtesy and concern for plantation owners as you do abolitionists."

"Abolitionists don't come to Cape Island."

"Exactly. They're too busy making speeches elsewhere!"

Harrah had never trained his daughter to show the exaggerated respect other parents demanded of their children. He prided himself on being a liberal father. But at this moment he regretted that he had never taken Willacassa in hand. And he realized it was too late to do anything about it now.

"I think you've said enough, young lady. Go about your business. I have things to do."

"And so have I!"

Not a friendly parting, observed Harrah, considering their closeness of the day before. Just as Francis Grandee's letter had brought them together, now it was driving them apart.

FROM THE PULPIT

For weeks Eliot Irons tried to persuade Willacassa to speak against slavery in one of Cape Island's churches. It was important to get the message out. For too long Cape Island had been sheltered from controversy. The issue had to be taken out of the side rooms and back-road cottages and brought out into the open—in the very heart of town.

"Why don't you make the speech, Eliot?"

"I wouldn't be as effective as you. For one thing, I don't come from Cape Island. I'm not known here. You are. For another, I'm not that good a spokesman."

"I don't know that I can do better."

"You'll make an excellent speaker. I honestly believe that. And I don't think being a woman will hurt either. Yours is a point of view people don't get to hear too often. And you're pretty to look at. That's a definite advantage—at least with the gentlemen."

So Willacassa decided to take the plunge. Though she was close to Reverend Moses Williamson, she did not ask for the Presbyterian Church at Cold Spring, which was three miles away. Nor did she request the Roman Catholic Church, which stood at the center of the Island, for fear of alienating the Protestants. She much preferred the "Visitors Church" on Washington Street, its pulpit being free to all denominations during the bathing season. And she was pleasantly surprised when the "Visitors Church" granted her use of the holy place.

Willacassa looked down from the pulpit at the rows of citizens who had come to listen. Actually only the first six or seven rows were filled to capacity: the rest were only sparsely populated. And though she recognized her father and some aquaintances among them, Willacassa was satisfied that for the most part she was dealing with the non-committed.

"It is difficult to believe," she began, not without her share of swallows and jitters, "that this is the Year of Our Lord 1859. For while men are free in the North, freedom is denied to Negroes in the South. I won't talk today about the on-going struggle of women, white women, for equality in both sections of our country. For, women at least are not in bondage. And a more pressing concern is salvation for the Negro.

"It is now several weeks since Harper's Ferry. I don't know if John Brown is the madman some people take him to be or 'that new saint,' as Ralph Waldo Emerson described him. But I do know that the barbaric, outrageous institution of slavery is still condoned in the South. We talk a great deal about this being a free country. And in some ways, for some people, it is free. But in the broad view of things no society is free that is also a slave society.

"Forget the cruelness of slavery, the inhumanity of it. Forget what it does to the Negro, if you must. But think for a moment what effect it has on you.

"No man of conscience can protest his own enslavement if he will tolerate the enslavement of others. 'As I will not be a slave, so I will not be a master,' said one of our countrymen a few years back. How wise is that sentiment! For a man who has a slave is as much a dependent and a burden as he claims the slave to be himself.

"We hear in the South the strident voice of superiority. 'God may have created all the races at one time. But he made the black race an inferior race.' The surprise is not that Southern planters believe this. The surprise is that the argument doesn't have one mule's leg to stand on. In fact, it is the Southern planter who is decidedly wanting. There is a certain mentality at work here that is not only deficient but incomplete. It is as if the Southern planter never grew up into the real world, much less the world of the Nineteenth Century. His parents had told him he was something special and, like all children, he wanted to believe this. And so he believed it and built a society on this romantic notion.

"But Southern aristocratic society is not a romantic society. It is a stupid society. It ignores the lessons of history. No people, however oppressed, are going to stay slaves for long. Somewhere along the line the Southern planter has come up with the notion that he is descended from the Romans, if not lineally, at least spiritually. The Romans had slaves. The Southern planter, then, must also have slaves. The Southern planter forgets that the Romans lived in the first four centuries A.D. We are now past the midway mark in the Nineteenth Century! Slavery is but an anachronism, passionately held onto by another anachronism, the Southern mind. Let's not humor this mentality even if Cape Island is the meeting place of North and South.

"Let us not perpetuate the myth of race superiority—at least from a source that is poisoned by the well-spring of greed. Slavery is evil and indefensible. Its destruction is a necessary good.

"Let us from this day forward proclaim the commonality of man. If we do this, we may yet escape the bloodbath that will otherwise engulf us all."

REACTION

The reception to Willacassa's speech was divided. On the one hand, there was Joseph Leach of the *Ocean Wave* who in an editorial pronounced Willacassa the Jeanne D'Arc of abolition.

"I have read the speeches of Lovejoy and Garrison and Wendell Phillips. They may have been more erudite and more fiery. But none have shown greater insight or greater conviction than that shown by our young lady from Trescott House. It is still not too late to prevent a holocaust in this nation, with section pitted against section, black against white, brother against brother. Miss Harrah's speech has crystallized the issue here on Cape Island. Let us hope it unites us rather than divides us, as we are already much too divided as a nation."

Others were not so impressed. Until now most of these individuals were not aware of Willacassa except as the pretty daughter of Nathaniel Harrah whom they regarded highly because of the quality and success of his small house. But this speech of Willacassa's had thrust her into the public consciousness of Cape Island, and she became the subject of gossip at every breakfast table and drawing room.

"She used to be such a sweet young thing. Remember how that young Southerner, Grandee, fell all over her."

"I remember how she broke off with him. He still comes to see her every summer. But she hasn't been the same since."

"She has another admirer now, a young man staying with the Wares."

"That's not the point," said one of the hotel owners to his wife. "I don't care about her suitors. The girl has singlehandedly disturbed the peace of the Island. We have our hotels to think of. If Cape Island becomes a hotbed of abolition, our Southern friends will go elsewhere. There's almost thirty years of tradition here. And it will all go out with the tide."

"What can we do?" asked a fellow owner.

"There's only one thing to do."

They took their case to Harrah.

Willacassa's father listened politely, but was unsympathetic.

"I rather liked the speech," he said. "And apparently Mr. Leach did too."

"Joseph Leach doesn't necessarily speak for Cape Island."

"His newspaper is widely respected."

"Forget Leach. It's your daughter we're worried about."

"All she is doing is speaking out against slavery. She has strong feelings on the subject."

"Look, Harrah, I don't hold with slavery either. But almost half our visitors—and by far the wealthier half—are from the South. Does it make sense to inflame their passions, to alienate them?"

"What do you want me to do?"

"She's your daughter. You must have some influence over her. Tell her she comes on too strong. Talks like that church talk are better suited for Philadelphia than Cape Island."

"I have no influence over her on these matters." admitted Harrah quietly.

"Well I don't mind telling you some people are riled up over Willacassa these past few years. There are even rumors that—"

"What kind of rumors?" Harrah's ire was now roused.

"Well, I don't know how true they are, mind you—"

"What kind of rumors!" Harrah repeated.

"Well, maybe rumors is the wrong word. Suggestions, let's say, that Willacassa has something to do with runaway slaves. Now I'm not saying that she's hiding them or anything. But some people seem to think that she's sympathetic to such goings-on. Until the last year or so Cape May County hasn't been involved with the Fugitive Slave Law—in any illegal way, I mean. But now there's people who claim to have seen runaways—at least to the west and north of here. I've even heard it said that Federal agents from Philadelphia have been to the Cape to look around."

"Where?"

"Near Fishing Creek Beach, near Town Bank, as close as Higbee's Beach. In '58 four blacks stole a government revenue boat in Delaware. The boat was found abandoned with all its equipment intact at Fishing Creek Beach. The four men disappeared. They were never found. Probably got help in Salem and made their way to Philadelphia."

"Or Canada," added another.

"What's that got to do with Willacassa?"

"I'm not sure. But the talk is she knows something about what happened."

"I think there's entirely too much talk," interjected Harrah angrily. "Maybe Willacassa shouldn't have made her speech at the church. I don't know. It's still a free country. But I won't have anyone on Cape Island criticizing her for it. If you want to let loose the mill race of gossip, there's plenty of things I know that would make interesting listening. A mite more damaging to the hotel business than a few disgruntled Southerners, I might add."

"What are you talking about?"

"A certain bank failure. A certain fire. A certain black mistress in the off season. But I don't put much stock in such things. And I suggest you don't give them currency either."

Willacassa met Eliot Irons at S. R. Ludlam's confectionery store, opposite the American House. Here Irons purchased two pounds of apples and Willacassa some cakes for Trescott House.

"That was my first and last speech," declared Willacassa.

"What do you mean?" asked Irons.

"I don't like making converts. If I have to assimilate people into the anti-slavery movement—if they aren't there already—then I have no use for them."

"But the message must get out," said Irons. "Especially where there's indifference. And those on Cape Island are as indifferent about slavery as any people I've met."

"That may be," said Willacassa. "But it suits their purpose to be sympathetic to the South. Besides, you've made speeches, haven't you?"

"Yes, in Boston, Camden, Philadelphia."

"What good did they do?"

"I like to think they did some good."

"Did all those speeches of Lovejoy and Garrison and the others win anyone over? No, the people that came to listen were already sympathetic to the cause. Or ready to throw stones. Only the martyrdom of Lovejoy galvanized support for the movement."

"So you're through with it?"

"Exactly. From now on I'll just do what I have to do. If they scream about it, let them scream. I'll let no one get in my way."

"As noble a sentiment as I've ever heard," chided Irons.

"Noble or not, that's the path I'll take."

RUNAWAY

"Where's Harriet?"

"Gone," said Harrah.

"Gone? Where?"

"I don't know." Harrah pulled up a chair. "Sit down, Willacassa."

Willacassa looked at her father who wore his rare no-nonsense look and decided it would be best to take his offer.

"Some federal agents came looking for her today. They wanted to know if a Harriet Tubman was working here. I said I didn't know a Tubman."

"But—"

"Well, that's not the last name she gave me. In any case, they looked around but didn't find her. I had sent her to Bickcom's Bakery."

Willacassa waited for her father to go on. When he did not, she asked, "Well, what happened?"

"She never returned."

"Do you think they got hold of her?"

"No."

"What makes you so sure?"

"The agents have been back looking for her."

"What did they want with her?" Willacassa asked this not so much for her own enlightenment as to learn how much her father knew.

"There's a price on her head. As much as ten thousand dollars."

"Ten thousand dollars!"

"They say she's been helping runaway slaves. A couple of hundred is the figure they gave me. She's even been known to point a pistol at the timid ones to keep them going."

"Good for her!"

Somewhat alarmed by the outburst, Harrah leaned over and whispered to his daughter.

"Listen to me, Willacassa. I don't know what you're involved in. And I don't want to know. But as your father I can only give you a word of caution. These federal agents play rough. They enforce the fugitive slave act to the letter. Particularly in New Jersey. Don't ask me why New Jersey. But it's a fact that while the law may be honored in the breach elsewhere it's totally enforced here."

"I don't know what you're talking about, Father."

"I'm saying the agents may be back. So don't do anything

rash."

"I have nothing to do with slaves. Or the fugitive slave law. So I have nothing to fear. But thank you anyway. Did Harriet leave anything behind?"

"Only some clothes. Nothing the agents wanted."

Willacassa surrendered the chair her father had given her.

"Do you think we will ever see Harriet again?"

"Not likely," said Harrah. "Apparently she has more important things to do than making beds in Trescott House."

1860

MATTERS GREAT AND SMALL

THE *GREAT EASTERN*

The *Great Eastern* was coming. Rumors had been flying for weeks, but with an editorial on July 26 the *Ocean Wave* made it official. The great "Leviathan," the Tenth Wonder, the largest ship in the world was coming to Cape May. Leaving New York City on July 30, it was expected at the Cape early the next day and would leave again that evening.

"This trip has been arranged with a view of affording excursionists an opportunity of witnessing in operation the vast motive power of the ship, both paddle and screw engines. Dodsworth's celebrated bands, both military and cotillion, will accompany the excursion. Tickets for the round trip are ten dollars."

Willacassa could not believe the excitement this stirred up. Cape Island was invaded by every form of vehicle: carriages, landaus, broughams, wagons, buggies—all converging on the resort as though making ready to embark in twos on Noah's Ark. The hotels and boarding houses, the cottages of the rich and poor, even outlying farmhouses were crowded to the wall. Cots and mattresses were thrown on the floors of vestibules, dining halls, and verandas. Visitors of every description, unable to find lodging, prepared to sleep on the beach or in backyards. Not a few horses had to share their stables with overnight guests. Even Eliot Irons came in from Salem.

"We have no room, Eliot," Willacassa teased him from behind the desk at Trescott House. "But I could let you sleep under the porch. It's quite dry when the tide is out."

"Anything will do."

"And what is your great interest in the ship?"

"You forget that my father was a sea captain. From the age of five I was raised in a ship's cabin. If he hadn't gone into the slave trade, I would have followed the sea."

Willacassa did not know how much of this was an excuse and how much sincere remorse.

"I'll find room for you on one condition," said Willacassa.

"And what is that?"

"You take me with you when you go aboard the *Great Eastern*."

"I was planning to take you anyway. Isn't your father going?"

"My father hates crowds and what he calls unseemly spectacles. It will be just the two of us."

"Fair enough. Now where will I sleep? In your room?"

Willacassa laughed. "Is this Eliot Irons talking? The holier-than-thou? No, you'll sleep in my father's office. We might be able to squeeze a cot in there. Can you believe how many people are in town?''

By weekend Willacassa realized that even she had underestimated the numbers thronging to Cape Island. On Saturday the steamboat *Delaware*, which plied the waters between Philadelphia and New York, spilled some three hundred passengers onto the boat landing. Later in the day the steamboat *George Washington* dropped off two hundred Philadelphians. On Monday the *John S. Shriver,* an iron-propeller ship, brought in excursions from Baltimore and Philadelphia. And another eight hundred curious were delivered by Captain Flowers and the *Kennebec*, its stack smoking and the huge paddlewheel churning. Dozens of others were floated in on small excursion boats from Chester, Wilmington, and New Castle.

And then the watch began. At dawn Tuesday morning Congress Hall, the New Atlantic, the Columbia House, Trescott House, and all the other hotels on the shore were emptied of their occupants. Men, women, and children walked down to the beach. Old spy glasses were pulled out. Cannon were loaded on the lawn of Congress Hall. Banners of every color and description were hauled up flag poles.

"Do you think it will get here on time?" asked Willacassa, chilled by the morning air.

"Unless the pilot oversleeps or the helmsman goes off course," replied a raspy Eliot Irons.

At about half-past six someone shouted, "There she is!"

And there she was, with huge paddle wheels, all 692 feet of her, all five funnels and six masts and some 65,000 square yards of canvas. Indeed a wonder as she approached! Indeed a Leviathan! Bigger than any hotel on Cape Island!

"My God, she's beautiful!" exclaimed Willacassa. "I can't wait until we board her."

Now the rush began. Hundreds started for the boat landing to meet her. The streets became impassable. And the turnpike was transformed into a solid two and a half miles of horses, carriages, and pedestrians.

"We should have gone on horseback," regretted Eliot Irons.

"Or by boat. I saw scores of steamboats, schooners, even rowboats heading out to meet her. They'll get there before we do." If she was not exasperated, Willacassa was at least impatient to reach the landing.

When Eliot Irons finally maneuvered his carriage near the Oscar House, Willacassa saw the *Great Eastern* at anchor in the bay a few miles off the steamboat landing. Small steamboats, schooners, and yachts moved in and out of the landing wharves bringing people to and from the huge ship. Eliot, having eaten no breakfast, wanted to get some refreshments at the Oscar House. But the structure was already crammed with people. Mrs. Little's Saloon, opposite it, was doing a thriving business.

"Let's go aboard," pleaded Willacassa. "We can get something to eat there."

"All right."

Irons knew someone on the *Washington* and for fifty cents each way arranged to have the steamer take them to the *Great Eastern*. When the boat reached the mammoth ship, it had to wait its turn for access to the single gangway to the boat. After tossing about for almost an hour in the water, which did Irons's empty stomach no good, the *Washington* succeeded in coming alongside. But neither the ship's officers or the ship's crew would take the *Washington's* rope. And Willacassa could hear the captain shouting that there were too many visitors on ship already.

"Take the rope!" shouted the *Washington's* captain, "or we'll stove a hole in your side!"

An officer aboard the *Great Eastern* laughed broadly at such a notion. But he caught the rope and, after securing it, helped the passengers up the gangway. At the second landing of the gangway, the passengers' reception was more cordial. A natty midshipman saluted them and directed them to the ticket office where a fifty-cents admission fee was collected.

If either Willacassa or Irons had any notion of a leisurely stroll on the broad decks of the *Great Eastern*, this idea was quickly abandoned. Caught in push of the crowd, they were carried forward in sometimes conflicting directions, finally coming to a stop at the giant wheelhouses. Mixing with the regular passengers, they cast a look at the distant shore. Here swarming crowds were gathered, waving handkerchiefs and flags, but looking for all the world like Lilliputians.

"It's terribly exciting!" gasped Willacassa, her hair flying in the wind and her crinolines braving the breeze. "The ship's passengers must be thrilled by the spectacle."

"I doubt the passengers are getting any pleasure out of it," remarked Irons. "They're being pushed and buffeted. And are as much objects of the curious as the ship is. They must see us as nothing better than South Sea Island savages coming off our canoes to pay them a visit. The least we could have done was bring them flowers."

"Oh, Eliot, you have no romance in you! I sometimes think

you're a minister's son rather than a sea captain's!''

"Am I to be scolded by you?''

"No, let's eat. Or look at the engines or whatever else catches your fancy.''

With more searching and milling around, they found a dining room. After considerable delay attempting to catch a waiter's eye, they were served a standard fare of cold meats, bread and butter, and cups of tea or coffee. Water was to be had at no price.

"If Trescott House served meals like this, we'd go out of business in no time.''

His hunger assuaged, Irons led the way to the long tunnel which ran from the screw to the paddlewheel engines.

"I'm told the screw propeller has four blades and is twenty-four feet in diameter. Imagine, twenty-four feet!''

"Fascinating!'' feigned Willacassa.

A peek into the boilerroom revealed ten boilers and a score of unshirted stokers.

"She has one hundred twelve furnaces in all,'' Irons was told by one of the stokers. "And she carries fourteen thousand tons of coal. She's a big one all right. There'll never be a bigger!''

"I'm getting hot and covered with soot,'' pleaded Willacassa. "Let's go back on deck.''

Reluctantly Irons agreed.

On the promenade deck Dodsworth's band struck a few tunes, but all attempts to clear a portion of the deck for a dance failed.

On the ship's rail, with gulls circling overhead, Irons suddenly turned romantic. "You're a good sport, Willacassa. You didn't have to go down to the boilerroom, but you did. I loved you for it.''

"Good. Then you'll buy me a new dress.''

"I'm serious, Willacassa.''

She gently removed his hand from hers. "I know you are, Eliot.'' She moved away from the rail to examine one of the lifeboats.

"Don't you think we should be leaving soon?'' she added. "Some of the passengers are returning from their stop at Cape Island. And the visitors are leaving.''

Leaning over the rail, she noticed that the boats bringing the passengers back were having a difficult time reaching the now-battered gangway. The last propeller ship to arrive crashed against the side of the *Great Eastern* with every rising wave. Each time a little more of the ship's gangway was crushed by the contact.

"By the time we leave there'll be nothing left to step off with,'' she observed.

As they waited for the *Washington* to approach the ship, both

Willacassa and Irons watched some passengers scurrying about for sleeping quarters. A few men appeared on deck with mattresses under their arms. Another carried a couple of camp stools and set them down against a wall. One brave soul uncovered a lifeboat and climbed in, tucking the canvas back over his head.

"You wouldn't want to stay?" asked Irons.

"And sail all the way to New York Harbor?"

"And beyond!" he nodded, seizing the moment and taking her hand.

Willacassa was thrilled with the idea. This was the kind of spontaneity she looked for in Eliot but did not always find. To think, a voyage to New York! She had never been to a real port, a great metropolis on the Atlantic, a sprawling city bustling with commerce and teeming with people from every corner of the world. And for a moment she was mightily tempted to take up his offer—even the offer of marriage that was implied by the word "beyond." But in the midst of her excitement, she struck a cautionary note. It dawned on Willacassa that she was not ready to let her happiness depend totally on one person, even so close and dear a person as Eliot. And to make the trip without some kind of commitment seemed out of place. She struggled for the right words, as Eliot with passion in his eyes pressed his case.

"I know that I have no profession to offer. I don't regard my surveying as anything but a part-time profession. But there is our work. And I do worship the deck you stand on. Even the air that quietly disturbs your crinoline. I love everything about you, Willacassa. And I've had this feeling for a long time. But until now I've been afraid to—"

She squeezed his hand, partly to halt his mad gallop and partly to show affection. "If I were to marry anyone now, I'd marry you, Eliot. You're as good a man as a woman could want. But it's I who am afraid."

"Afraid?"

"I'm afraid that in some ways I am not the right woman for you. I hope I'm wrong. And time may prove me so. But if I do agree to marry you, I want to be sure. You understand, don't you?"

The fire that had been in Eliot's eyes suddenly blew out. Not even its embers remained.

"Of course." He turned towards the all-but-destroyed gangway. "If we are going to abandon ship," he said quietly, "I guess we'd better leave now."

SELF-DOUBT

Willacassa stopped by at Joseph Ware's house.

"Is Mr. Irons at home? We haven't seen him in several days."

Mr. Ware nodded.

"Where is he?"

"Upstairs. Sleeping it off."

"Sleeping it off?"

"Yes. Generally, Mr. Irons is a sober man. But every so often he brings some liquor to his room and makes a night of it. When we hear the crashing of bottles, we know it's all over."

This was stunning news to Willacassa, all the more because it was so unlike her perception of Eliot.

"We allow him to stay on because it doesn't happen too often. That and the rent are the main considerations. Mr. Irons is here most of the year now. Which makes a considerable difference in income for us."

Willacassa was twice humiliated. For herself as Eliot's friend and for the reputation he was undoubtedly making for himself. But she was not about to run away from an awkward situation.

"May I see him?"

Ware thought a moment, then nodded.

"Let me accompany you upstairs. He may not welcome a visit just now."

At the top of the landing, Mr. Ware knocked on a bedroom door.

"Mr. Irons, there's a visitor to see you."

When there was no response, he knocked again. "Mr. Irons, someone to see you, Sir."

Mr. Ware was about to knock a third time when a weak voice replied. "No one. I don't want to see anyone, Mr. Ware."

"For God's Sake, open up, Eliot! It's me, Willacassa!"

The hush that followed was painful in its forlorn solitude.

"Did you hear me, Eliot? It's Willacassa."

"Go away, Willacassa. I can't see you now. Give me a few hours and I'll meet you later."

"If you don't open up now, Eliot, you'll never see me again!"

Willacassa did not know if she could back up her threat or even if she wanted to. But she was mightily relieved when Eliot opened the door and Mr. Ware returned downstairs.

"You might have spared me this, Willacassa," said Eliot, looking as shamefaced as a churchplate thief. But he stood aside to let her in.

A brief inspection of his room confirmed what Mr. Ware had said. Broken bottles and shards of glass were strewn about the floor. Above the rumpled, unmade bed wine stains were splashed upon the wall. And the one portrait of a four-masted schooner, billowing in a storm, hung askew on the opposite wall and was in danger of falling from its hook.

"Well, now you know," said Irons, reading her mind as easily as his own.

"I had no idea, Eliot." She could say no more.

"I suppose until now you've thought of me as a sober, bible-thumping New England abolitionist," said Irons in a display of self-mortification. "Well, I'm not always sober. And I don't make particularly good speeches. Worse still, I don't write half as well as Wendell Phillips. And I hate myself for it. Because if I could only find the right words, I'd be able to turn hearts and heads away from this abomination of ours."

"You're doing what you can, Eliot. We're all doing what we can."

"But it's so piecemeal," he groaned. "It's like scooping sand with a spoon. How will we ever get rid of it?"

"How? By searching—relentlessly searching for a way."

"Searching? That's all I do. And it's all in vain. Sometimes I wish I was born a hundred years hence. Maybe by then the problem will be solved."

"Long before that," said Willacassa, "with any luck at all."

"Meanwhile," moaned Irons, "I can't bear the guilt. It's like the Scarlet Letter 'A'."

Willacassa had a faint recollection of Hawthorne's novel. But what did adultery have to do with Eliot?

"Only in my case," continued Irons, "it's the black letter 'S' that's emblazoned on my chest. Do you know that when I lie on my bed at night, I think I can see it? It's only when I light the lamp that I'm convinced the letter's not there."

Eliot Irons's face was squeezed with the pain of his torment. Gradually the handsome, regular features surrendered to a sea of crevices. When he could no longer fight back the tears, he looked frantically for help and Willacassa stretched out her arms towards him. Falling into her embrace, he buried his head in her bosom.

"I feel so inadequate," he cried.

"But you're a brave man, Eliot," said Willacassa, experiencing a small contempt even as she tried to console him. "And probably the best educated, intelligent man I've ever met. Why, it was you introduced me to Hinton Helper and Frederick Douglass and William Lloyd Garrison."

"Intelligent, yes. Intelligent enough to appreciate other men's talent, other men's genius," he said, looking at her once more as she held him at a distance. "But not intelligent enough to create persuasive arguments—to find words so eloquent and incisive that other men soon drop their opposition to what I stand for."

"Men aren't persuaded by arguments, Eliot. Surely you must know that. It's self-interest moves them. No argument, however true or brilliantly expressed, has ever triumphed over self-interest. It's only those with no vested interest who can be persuaded."

"And I'm not sure I can persuade even them. That's why I've taken to drink when I've never drunk before. That's why—"

"That's why what, Eliot?"

"That's why I may leave Cape Island."

"Leave? But this is the time of year you can be most useful. Hannah Warren—"

"It's not Hannah Warren, Willacassa. It's you."

"Me?"

He tried to face her squarely.

"Yes, it's bad enough knowing I'm not adequate to the job. Knowing you're indifferent to me makes it all the more unbearable. You must know I love you. And you know better than I how futile that love is."

Willacassa refused to let him exploit the moment. "I know only that you're feeling sorry for yourself, Eliot. And that's a luxury those in our line of work cannot afford. Now wash up and meet me in an hour. And maybe we'll have lunch together."

As she was getting ready to leave, Willacassa noticed a picture on the bureau near the door, which had escaped the general destruction.

"Why, Eliot! What a pretty woman! Who is she?"

"It's a likeness of Ann Terry Greene, made some twenty years ago."

"The wife of Wendell Phillips?"

He nodded sheepishly.

"But I don't understand. Are you enamored of her? She must be past forty by now. Is she another of your older women?"

"No," he said simply. "But I've always admired the kind of woman she is. Supportive, steadfast—the ideal partner for someone like Wendell Phillips."

As Willacassa remembered it, Wendell Phillips was a tall, handsome man of superb physique—not unlike Eliot in appearance. And his wife, the daughter of a wealthy Boston merchant. Reportedly she was an invalid in past years, confined to her room and often to her bed. Still, she was the acknowledged force behind her husband's crusade and the source of their very happy domestic life.

But what did all this mean? And why the picture, almost shrine-like, on Eliot's bureau? She did not have to wait long to find an answer. Eliot himself provided the clue.

"What distinguishes an Eliot Irons from a Wendell Phillips? I keep asking myself that, but fail to find an answer. Is it schooling? I'm as well educated as he is. Is it conviction? I believe with all my heart that slavery is wrong and would as soon see the Union dissolved as see it half-slave. Is it determination?"

He looked at Willacassa. "Do I lack determination? Is that the missing ingredient in the soup of my personality? Perhaps if I were more determined, I'd sweep you off your feet. That's what women want, isn't it?"

He turned away from Willacassa and looked at the picture of Ann Terry Greene.

That was part of it, acknowledged Willacassa to herself. The other part, she suspected, had nothing to do with determination or ability. It had to do with being at a critical place at a critical time. Had Eliot Irons been in Boston during the attack on William Lloyd Garrison by a mob of "respectable" citizens, it might have been he who had the incident branded into his soul instead of Wendell Phillips. Just as the sight of half-a-dozen chained black men lying dead on Higbee's Beach had been branded into her consciousness these many years.

But as she considered it further, Willacassa realized that it wasn't by chance that Eliot always conjured up the name of Wendell Phillips and not that of Garrison or Lovejoy. The reason for his alluding to Wendell Phillips was his wife. In 1835 the young and brilliant former Ann Terry Greene prodded her husband into making a total commitment to the anti-slavery movement. Willacassa did not know how much prodding this had required. But this remarkable young woman had somehow enlisted Wendell Phillips's enormous energies and writing skills whole-heartedly in behalf of the cause. Could Eliot perhaps have seen in Willacassa another Ann Terry Greene, a young and pretty woman, a future wife, who would prod him into greatness?

She hoped not. For if that's what Eliot envisioned, he had greatly miscalculated. Willacassa would prod no one. Only herself. As fond as she was of Eliot, she would be neither prod nor spur for any man.

"I wish I knew what women wanted, Eliot," Willacassa responded at long last. When she left, she barely whispered goodbye.

SOMETHING TO BE SAID FOR FRIENDSHIP

"Do you think I'm becoming a spinster?"

"No," said Harrah.

"But I'm near twenty-three and not married."

"Your mother was more than that when I married her."

"But you lived together for a while."

"How did you know that?"

"Rachel told me."

"Rachel? I thought all she did was tend to her cooking."

"She talks on occasion—when I pull it out of her."

"I'll have to speak to the woman."

"It won't do any good," smiled Willacassa. "She won't talk to you. After all these years she's still afraid of you."

"Afraid?"

"Well, in awe of you. Your having worked for the government and all."

"I can't believe that."

"It's true. I wish I knew you then," said Willacassa thoughtfully.

"Why?"

"Maybe I'd know what to look for in a would-be husband."

"What about Eliot Irons?"

"I'm extremely fond of him. I admit that. And I've learned a lot from him about the world outside. But he's not spontaneous enough for me—although he does surprise on occasion."

"Still, he's a fine man. And strong in his way."

"I don't argue that, Father. But when I look at Eliot I sometimes feel—"

"Feel what?"

"That he's doomed."

"Doomed?"

"Yes, just as the nation is doomed. I know it sounds strange. But there are times he has the look of death. It blanches his face and his eyes. I get the feeling that one day he will become the casualty of his own private war. It's as though the world is too much for him."

"Aren't you being unjust?"

"I suppose I am. But I can't escape the feeling. It's as if he, not the President, has the cares of the nation on his shoulders. It's as if the conflicts between North and South are tearing him asunder instead of the Union."

"Do you miss Francis Grandee?"

"I miss the fun of Francis, the boyishness. Though I have nothing but contempt for the boy he is and what he stands for. Do I sound like a terrible person?"

"No, Willacassa. You sound very much the young woman I had hoped you'd become. Frankly I'm proud that you broke off with Francis. And I'm glad that you're not rushing into marriage the way some young women do. You and Eliot can always be friends—even if he's not the husband for you. There's something to be said for friendship."

"Friends like you and Madeleine Culpepper—even though she remarried?"

"How did you know about that?"

"Mr. Leaming told me. And I've seen her letters for you in the Post Office."

"Maybe you ought to be the town gossip instead of somebody's wife," chided Harrah, leaning over to kiss her. "You seem to have a talent for it."

Willacassa was right. Madeleine Culpepper—her name was Craddishaw now—and Harrah had become friends. For the past three years she had been writing to him once a month, with such regularity that Harrah liked to think she wrote when her time had come, as a kind of relief from tensions. But he suspected that he was being unkind and that whatever seeds of companionship thay had planted during their time together on Cape Island had borne fruit.

It had been a bitter pill at first, accepting the idea that their love affair was over. How could such intensity of passion and intimate friendship lose their sting? Harrah was still as much in love with Madeleine Culpepper as he had ever been when she graced Cape Island. Even now it pained him to think of her, of their time together. And the pain of that love, however dull the passing of time had tried to render it, was still piercing. Sometimes Harrah thought there would, through the alchemy of undying devotion, be a reestablishment of their relationship. But the fact of the situation— that she was there and he was on Cape Island— confirmed his suspicion that he was deceiving himself. To tell the truth, there was precious little he could do to change things He was reduced to being a pen-and-ink lover on half-sheets of fool'scap, while she had both a husband and, now by virtue of the postmaster, a companion.

Madeleine proved in fact to be an excellent writer, and the mail he received from her could aptly be described as "belles lettres." At first she wrote of her life as the wife of a Virginia aristocrat, at least the ordinary details of it, which were no longer ordinary but revealing when Madeleine gave them her finishing touch. She had a ready wit, penetrating but not acerbic, which she used to good advantage especially when she described the social give-and-take at balls and parties or the habitual malaise of those confronted with the challenge of work or original thought.

But she was at her best when she painted a picture of the South, at least Virginia, struggling to preserve both the aristocracy and its "peculiar institution."

"I won't say these men are plotting secession. But in their vanity of vanities they are creating a climate for secession. They seem to think—and rightly so—that a way of life, their way of life, is hanging in the balance. They see slavery not only as an institution, but as a form of free prostitution. You can't believe how lecherous some of these distinguished statesmen are. Oh, they are perfect gentlemen with the ladies assembled at their dinners and parties. These ladies are, of course, of their own race. Occasionally they will engage in flirtations. And I am sure one or two of them have white mistresses, kept at great cost. But the majority of our planter aristocracy have their own black harems, choosing the sweetest and tastiest of the Negro girls for their own. They would no more see slavery abolished than they would tolerate an end to fox hunting or drinking or gambling!

"My husband, of course, is the rare exception—though why he hasn't strayed I do not know. For I am cold to him at times and I have never been passionate by nature. He says I husband all my warmth and ardor for motherhood. This may be true. My children do not complain of my excesses, however. Only Benjamin complains.

"But I am sure you don't understand any of this. I wonder what kind of a woman you think me when you read my letters. But that is unimportant now, isn't it?"

Harrah did not know what to make of her letters. Madeleine wanted no doubt to reprove him. She had wanted him as her husband, to make a life together in the ancestral habitat of her children. But the Commonwealth of Virginia—or was it slavery?—came between them. And in the end, much as he loved her, Harrah could not meet her terms.

And so she married another. See, she was saying, I made a life without you. It had its drawbacks, at least in her relationship with her husband, but she was playing the mother role to the hilt. A girl graduated from young woman to lover, to spouse, to mother. Each stage in life had its moments. And if a woman was able to add friendship to her list of accomplishments, so much the better, even if the friendship was at some distance.

But friendship was a two-way street. It was one thing for Madeleine to parade her thoughts and emotions. Harrah had thoughts to convey too. And if they could never ride together in romantic intimacy, they could at least maintain the illusion of close contact.

And someday perhaps—but there was no someday. That too was an illusion.

1861

WILLACASSA

RAILROADIN'

When Willacassa first heard that a boatload of blacks had been sighted, not far from the strand fronting the New Atlantic, she propelled herself into action. The runaways had apparently been swept off their course. Usually landings were made at night north of Higbee's Beach. That the blacks had been seen in broad daylight indicated that something had gone wrong or that an unscheduled stop was being made on the Underground Railroad.

There was no time to notify Hannah Warren. And Willacassa didn't have the faintest idea where Eliot Irons had been in the past few days. She therefore took it upon herself to find the slaves before anyone else did and to deliver them to the next station on the underground road to freedom.

Hitching a horse to her father's wagon, she lumbered up Ocean Street and swung into Washington Avenue. Higbee's Beach! Though she dreaded the place, she would drive there. It was a certainty the blacks would beach their boat near the Old Landing and scamper into the thicket.

And that's where she found them, the four huddling together in what they deemed to be an out-of-the-way place but one that could easily be seen from the Bay.

"Come with me." she announced, using every gesture at her command to prevent panic. "I'll take you to a safe place."

The blacks hesitated at first. But when Willacassa repeated her request, softly this time, with a gentleness in her voice that took the sting out of her sense of urgency, they left their hiding place. The men were obviously field hands, well-built and coarse-featured. One carried a gash on his right shoulder which had formed a rough, raw scab. Another wore a hand-made bandage on his wrist. The girl was the only one unmarked, and she was addressed as "Sissy" by the third male, the youngest and shortest of the group, who kept pressing his fingers to a raised welt on his forehead.

"You should know," the girl volunteered. "We had a scuffle before gettin' away. Had to strike a deputy with a shovel. Still want to help us?"

"Come," said Willacassa. "I'll show you where my wagon is." As an afterthought she added, "I can't blame you for fighting back."

The girl, a pretty thing with fine white teeth, smiled appreciation. Then she followed the big man named Jethroe as he trailed the sandy path taken by Willacassa.

It was fast settling dark. Helping the four into her wagon, Willacassa covered them over with canvas. Climbing aboard the driver's seat, she made for Hannah Warren's house. Not long after, she stood on the widow's porch and banged on the window. Hannah Warren only half opened the door.

"I have some 'passengers,' Hannah. Can I leave them with you until I make contact in Salem?"

"I can't take them tonight," the widow said hurriedly. "Federal agents were here. They'll be watching."

"But where can they go?"

"Try your father's cottage. But use the woods."

"My father's cottage?"

Hannah Warren was dumbfaced. "Didn't you know? Didn't he tell you? My poor girl." She tried not to give anything away, but involuntarily she gestured. "You can't see it from here, but it's the next house up the road. There's a secret room upstairs behind the cupboard. Hide them there. But don't stay long. Just until they're rested. The agents will be back, I tell you. They're checking every road, every house. If they pursued criminals the way they dog these poor people—but never mind, just be careful."

Willacassa returned to the woods where the four were waiting.

"Wrong place," she said, trying to calm their disappointment. "It's a little farther on. Then you can rest."

"How much futhah?"

"Just a bit more. Follow me."

Willacassa's head was swimming. Her father's cottage? Why didn't he tell her about it? Was it an alternate stop in the underground railroad? Stations and station masters were being changed all the time. Hannah Warren didn't have time to explain. But this was the way it seemed. Oh, how wrong she had been about him all these years! How deceitful her father had been. But apparently it was a glorious deceit, a deceit made at great personal sacrifice. How would she ever make it up to him?

The runaways were clearly exhausted, so she quickened her pace before they gave out, picking her way through the trees until she saw the shadow of the house across the road. Isolated the place was, but not so secluded as Hannah Warren's. On such a moonlit night anyone could see them crossing the road. But they had to chance it.

She stepped out onto the road which had always seemed narrow before but now appeared broad enough to make Willacassa feel conspicuous. Finding nothing in either direction, she signalled for her charges to follow.

Willacassa was not frightened. After all, it wasn't she who was being pursued. She was only helping the runaways to escape. But the blacks did not know where they were being taken, who would be there to meet them, or what the next house, the next day would hold for them. And so she understood why they looked so pale when there was sufficient light to reflect on their faces, why they trembled at the slightest sound or took false steps, why they seemed totally disoriented as to time or place.

She found the door to the house opened and ushered the runaways in. Stumbling upon a lamp on the living room table, she lit it. She did not bother to look around, to see what her father's cottage was like, something she would have been keen to do under different circumstances. Instead she led the three men and the girl up the narrow staircase.

In the bedroom there was no "cupboard" that she could see. But there was a bookcase which the men quickly pulled away from the wall. When three of the runaways entered the hidden room, Willacassa detained the man who appeared to be their leader.

"Look, Jethroe, we don't have much time. When we get to the beach, you'll find an ashpit big enough for a lobster pot. It is open at one end. If you face the pit at the opening, you will see a grove of fairly tall trees. There are a couple of overturned rowboats hidden there. One of them has a portable sail."

"You'll show us, won't you, Ma'am?"

"Of course I'll show you. But if we should be separated, you must go on without me. Take to the bay till daybreak, then get back on shore again and hide until nightfall."

Jethroe nodded at almost every word.

"In time you will pass two big creeks. When you get to the third, take the creek to Goshen. There's a lot of shipbuilding going on, so you'll know the place. Ask for one of the Sheppards. Any one of them will take you to a 'station' in Cape May Court House. That's the county seat. There you'll find another 'conductor,' like me. And he'll get you to Salem."

Again Jethroe nodded, not once taking his eyes from her.

"Use the papers I gave you. They say you're free. And if you and your friends believe it, anyone who asks you will believe it. Do you know what to do?"

"I remember every word you said."

"Good. Then get some rest. In a couple of hours, we'll be starting again."

She had just lain down on the horsehair mattress upstairs when she realized that she had left the wagon in front of Hannah Warren's house.

How could she be so stupid?

Of course it would have been no wiser to have left it standing in front of her father's cottage. She might just as well put up a road sign for the Federal agents to see. No, she would have to recover the wagon and hide it somewhere. She returned downstairs.

Willacassa did not go by the woods this time. She simply walked down the road about a quarter of a mile when she spotted the wagon standing almost luminescent in the bright night.

Her horse, which had not been tethered, had pulled the vehicle slightly off the road. He was contentedly munching away at the grass when Willacassa gathered up the reins. Climbing aboard, she clacked the horse into an amble. About halfway towards her father's place, she changed directions and drove the wagon towards the trees.

"That won't be necessary, Miss," said a horseman darting out of almost the same spot Willacassa had chosen for her hiding place. Three other horsemen slipped out of the woods and one of them, drawing abreast of the wagon, reached over and relieved Willacassa of the reins.

STIRRED TO ACTION

Willacassa's arrest stunned Cape Island. What peace and tranquility the community knew in the off-season turned to outrage. It was not that the Cape Islanders were unaware of the presence of fugitives on their shore. It was not that they did not know these runaways were being aided and abetted by local inhabitants. One had only to read the newspapers and the paid advertisements with offers for rewards to see that this was a commonplace occurrence. It was not even that one of their own was part of the illicit traffic. What disturbed them all was that she had allowed herself to be caught.

So long as Willacassa—or anyone else, for that matter—had conducted herself with impunity, they had no complaint. When the summer tourists arrived, no one would be the wiser—least of all those from the South. The shameful part was in exposure of the matter. Now the whole world, at least the world of Cape Island, knew that what had only been rumored before was indisputable fact. Willacassa Harrah, daughter of a prominent innkeeper, was a conductor in the Underground Railroad. Even her abolitionist tirade in church was small annoyance compared to this. Cape Island's reputation thrived as a watering place, a peaceful coming together of North and South. It could only suffer from the partisan bitterness of sectionalism.

The effect of Willacassa's arrest, stunning as it was to the rest of Cape Island, was absolutely devastating to her father. Normally cool and careful of demeanor, Harrah found himself consumed with anger. Where before he had held the idea, the prevailing sentiment, that the anti-slavery movement would require time and patience if it was to attain its goal, he now saw his daughter's detention as symptomatic of what was wrong with his countrymen's attitude. It was too pacific, it was too legalistic, it was too rational. Mayhem was being committed. Human bondage was being suffered. A gross injustice and immorality was being witnessed. Yet it was breakfast as usual in the houses and cottages of Cape Island.

Harrah blamed himself as much as any other man for this mood of normalcy in the face of such abnormality. So long as he had been reasonably comfortable in his own situation it was easy to wait out the storm. So long as Rachel placed eggs and a rasher of bacon on his table in the morning, the sweat and toil of black slaves, the separation of families, need not trouble him till noon or the arrival by steamboat of the Philadelphia and Camden newspapers. Harrah had taken the prudent, conservative route in radical times, even allowing for his own clandestine role in the Underground Railroad. In the final analysis, he had met inexcusable oppression with thoughtful opposition.

But no more.

With Willacassa's arrest, Harrah quickly moved into action. He first stopped at Joseph Ware's house and was told the long-time mayor had gone to Washington to visit relatives, but Harrah suspected the visit was part of Ware's perennial search to secure himself an appointment as Postmaster.

Visits to other city officials proved to be no more productive. The off-season, especially early spring, was a time of absence from Cape Island, a last chance to find diversion before the peak demands of summer. And everyone from the alderman to the councilmen to the recorder were either out of town or unwilling to take any action in the case.

Only the Constable, Emory Thornton, assumed any authority in the town. But authority was to Thornton what drink was to John Barleycorn. Indeed, Thornton exemplified the new officiousness that had overtaken Cape Island since it became a municipality. Prior to 1851, there were local ordinances but no one paid them much heed. The hotelkeepers assumed responsibility for the peace and quiet of their houses and the carriage drivers for the street. After 1851 the constable took his job seriously. He enforced the ordinances and he brought "sharpers" and pickpockets before the mayor to be fined or jailed as befit their crimes. And in the past decade stringent measures were adopted to ensure proper decorum and respectable behavior in the resort. These too, for the most part, were enforced; and in recent years Thornton was the enforcer.

And it was Thornton who insisted Willacassa be held either until picked up by the Federal authorities who had gone to Philadelphia to secure the necessary documents or until he had instructions to deliver her to the Federal District in Philadelphia.

"You're not serious, are you? You've known Willacassa ever since she was a schoolgirl at Cold Spring Academy."

"I'm always serious, Mr. Harrah."

Thornton was a thick-set man, with a thick neck for his thick head to rest on. He was not an ugly man, but he was so humorless and such a stickler for the law that Harrah found him insufferable to be with.

"But no other state enforces the fugitive slave act."

"So' Jersey does. And I was sworn to uphold the law," said Thornton drily.

"The least you can do is wait until Mr. Ware returns."

"Neither Mr. Ware nor anyone else can deter me from my

duty.''

They were joined by Joseph Leach.

"You know there's no reward in Jersey for turning in fugitive slaves," said the editor of the *Ocean Wave*. "Or those that help them."

"I know," said Thornton who never liked Leach anyway.

"Then what are you doing this for?"

"It's the law of the land.''

"Did the men have a warrant for the fugitives' arrest?"

"I don't know."

"It's a misdemeanor to seize or remove alleged runaways without the necessary warrants."

"Are you a lawyer, Mr. Leach?"

"No. But I have been a justice of the peace."

Thornton rose from the small wooden table that served as his desk and lit his pipe.

"I'll worry about the misdemeanors when the time comes."

"Certainly you're not pro-slavery, Mr. Thornton. No true Jersey man is."

"I'm not for and not against. Like most Cape Islanders. But I'm not for freeing them either—when they're another man's property."

Joseph Leach put his hand to his brow. "My God, man, don't you know this is the Nineteenth Century! There are over 25,000 free blacks in New Jersey alone!"

'There are slaves, too, I hear.''

"Only 18 by last count."

"Well, all that is besides the point," said Thornton. "The fugitives were never caught."

"Never caught?"

"No, the agents searched all the houses in the area. No sign of them. Must have escaped along the bay."

"Then why is Willacassa being held?" asked Harrah.

"She admitted helping them. Said she was glad they got away."

Just like Willacassa, thought Harrah, with a mixture of annoyance and pride at his daughter's action.

"Are you going to release my daughter or not?" he asked, beginning to lose patience with Thornton, with the slave act, with the whole sick business.

Thornton turned to the owner of Trescott House.

"I'll no more release her than I'd release Mary Dugan's slut if she were aidin' and abettin' black fugitives. It makes no difference that Willacassa is the daughter of a fine, upstandin' guest-house owner like you."

Harrah's anger turned to ice.

"Come on, Mr. Leach," he said. "We're wasting our time here."

He stopped at the jail door.

"Just one thing, Thornton. If you transport my daughter from Cape Island. Or if any of your deputies transport her, I'll knock your head clean off your neck. Is that clear?"

"I don't take kindly to threats, Mr. Harrah."

"Come, Nathaniel," said Joseph Leach. "You'll only make things worse for us."

"It's dreadful seeing you in jail, Willacassa." Francis Grandee was just as passionate in his outrage at seeing her in a dingy two-celled structure as he had been in his more romantic moments with her.

"You should be here, not me." replied Willacassa.

Francis Grandee was stunned by the remark.

"What do you mean?"

"For condoning slavery in the first place."

Francis Grandee winced at what he regarded as a betrayal of his new personality. After all, he had renounced slavery, an act which in the South was tantamount to sedition. He had kept his bargain with Noah, and then some. But he was determined not to quarrel with Willacassa.

"I've tried everything, even bribery, to get you out of here."

She softened but not too much.

"Yes, I know you've tried. Just as father has. And Eliot." She turned away from him. "And I know you're a changed man and that you mean well—now. But it's no use. I'll never forgive you for being what you were."

"A Southerner?"

"No, there are many men and women of good conscience in the South. They may not have much sway, but they're there. I'll never forgive you for thinking you're a privileged race. And for treating the rest like dirt."

Francis tried to hold his tongue but could not. "All right, slavery is wrong," he conceded. "But even you will admit the Negro is inferior to the white man. Anyone who has anything to do with blacks knows they're lazy. They'll only work when they're pushed to it. You talk about setting them free. And I'm in agreement with you. But if they won't work, how can they survive? How can they remain free? Until now they've got us to thank for surviving at all."

"I don't know why I even bother talking to you," said a dejected and weary Willacassa. "You're a lost cause—with or without your transformation."

Francis reached for her hand to try to stay the argument, but Willacassa pulled away.

"You can't help it , I guess. The notion's been ingrained from birth. You plantation owners think you're the saviors of civilization. So everyone has to cater to you, sweat for you, even slave for you. The fact is, you're the lazy ones. The ones that are uncivilized and violent—and rapacious. Not your slaves."

"I never raped anyone," protested Francis, divining her meaning. "That girl was willing!"

"But she was pregnant!"

"Makes no difference. She wanted me, I tell you. It was Noah put her up to complaining. Noah made her come to my parents."

Willacassa was horrified. Would she ever understand him? Would she ever plumb the depths of his degradation? "And that makes it all right, Francis! That makes it less appalling?"

"It makes it no different from anyone else. Why, your Northern men do the same thing when they come to the South to live! Or marry into plantation families! I know that for a fact. Don't make out that they're holier than we are!"

Willacassa despaired of going on. But she was so enraged that she could not hold back.

"I used to think you were a handsome man, Francis. Too handsome to be real. How I fluttered when I saw you! But with all this talk I see you differently. So much so that it defies explanation. The truth is, I can't stand to look at you anymore!"

There was no hiding the color in Grandee's face, nor the shock. Though he knew better, though he never doubted himself, he felt diminshed by what Willacassa had said. He could barely swallow, much less talk. But he managed to stay in control of his emotions.

"I know you're trying to hurt me." he finally blurted out. "And I guess you have. But no matter what you say to humiliate me, Willacassa, no matter what you do, I'll stand by you."

"I don't want you standing by me, Francis. I want you to go home. I want you to secede with the rest of South Carolina! I honestly think the Union will be better off without you. But free the slaves first. Until those slaves are free, you'll find no freedom either, only bloodshed. Mark my words!"

When she saw the pain on his face, the pain she alone could inflict on him, Willacassa felt a little better. With all deliberateness she watched young Grandee turn away from her and leave. It was only when Francis was gone and the jail door shut again that she hated herself for hurting him.

LOOKING FOR ANSWERS

Harrah walked the beach all morning. And all morning he heard the rough waves booming upon the shore. Like heavy cannon they thundered, spraying the sand with a watery fusillade. It seemed to him the sound of war was in the air; dull, thudding booms reverberating across the Island. And he was sure the puffs of cloud in the Cape May sky spelled the gray smoke of spent cannon fire.

As far as Harrah was concerned, it was war. Once his daughter had been caught in the crossfire of the fugitive slave law, there would be an all-out struggle to save her. And the incessant booming of the waves and the throbbing echo of thunder merely confirmed this. He could foresee no letup in the pounding. Until either the slaves were freed or the opposition destroyed, the booming at the shore would continue its dune-shaking barrage.

Why then had he remained neutral all these years? Why had he allowed himself to be lulled into a false political stance? Harrah tormented himself with these hard questions. It was, of course, false to assume a position that offended neither side. Even after he had allowed his cottage to be used by Hannah Warren as an emergency stop on the Underground Railroad, Harrah was still being polite to those who defended slavery "to the death" and whose every third word was "nigger" or "nigger-lover."

No, he should never have remained neutral. He should have shut the doors of Trescott House to them. He should have catered to Northerners only or those Southerners who, like Madeleine Culpepper, shared an abhorrence of slavery. But even as he thought this, Harrah realized that the summer visitors from the North were as much divided on this issue as the abolitionists and the slaveholders. How then could he have maintained a "Free Soil" house without making a fool of himself?

What Harrah should have done all these years was speak his mind. If he found something a patron said to be offensive, he should have noted this. And if the guest packed up his bags and stormed out of his house, with segar puffing, so much the better. A proprietor could live with an occasional defection. Indeed, such an encounter might have done the house honor.

Had he done this, Willacassa without a doubt would have had more respect for him. She would have seen her father not as an unobtrusive innkeeper, a public figure of no political persuasion. Rather she would have regarded him as a man of uncompromising principle, given as much to moral outrage as to making money from his hostelry. And if she had seen him in this light, there would have been no need for his daughter to go off by herself. No need to compensate for what she saw as the weakness of neutrality. No need to be a conductor in the Underground Railroad, derailed by the Fugitive Slave Act, arrested and now beyond his reach.

He took his case to Joseph Leach, Joseph Leach who had called for militancy while Harrah walked the path of patience.

"Look, Joseph, I know you have been far more active in the movement than I have. And I know that until now I have been dragging my feet. I recognize all that. But, for God's Sake, why don't we throw caution to the wind? Now that my eyes have been opened. Now that Willacassa is waiting to be transported to Philadelphia, I'm not satisfied with editorials anymore. Or speeches. Or promises of future undertakings. I want action. Meaningful, decisive action. Am I being unfair or unreasonable? Or just plain irresponsible?"

Harrah had uttered all this in response to Joseph Leach's reproach when the newspaper editor pushed aside his pen and ink bottle and said. " Beware of the convert. No passion is like his passion. No bitterness like his sense of outrage!"

"All right," continued Harrah, "you can ask, 'Where were you five, even ten years ago? Where were you when they passed the fugitive slave law? Or when they hung John Brown at Harper's Ferry?' And I have no answer. But now I see things differently, and I do have an answer."

"And what's that?"

"Action! Violence if necessary. I'll even kill if I have to. I'll kill to stop them from taking her!"

Leach sprang out of his chair and took his friend by the arm.

"Hold on, Nathaniel. You're no more commited to the slavery question now than you were before. The thing that really irks you is that your daughter was arrested. That's what it's all about, isn't it? Would you have been so incensed had Hannah Warren been arrested? Or Eliot Irons? Or me? No, you'd have shaken your head and said, 'What a damn shame!' Maybe written a letter. But no more than that. I say this even though you love your daughter as much as you do. You'd have been a man of peace in the middle of an undeclared war!"

Harrah was momentarily subdued by the truth of Leach's observation. The man was right on target, he had to give him that. But Harrah's new surge of anger was not to be denied. He wanted to strike out, to punish, to maim if necessary. And he vented his anger against the Constable of Cape Island.

"I'll knock his damn head off!" threatened Harrah. "If he doesn't let Willacassa go, I'll smash his stupid face in!"

"Thornton?"

"Yes, Thornton!"

"That will do no good, Nathaniel. You'll only find yourself sitting in the same jail—with a far stiffer sentence than Willacassa's staring you in the face."

"I don't care!"

"You'd better care. For Willacassa's sake."

Hannah Warren was of no more comfort than Joseph Leach.

"No, the Railroad doesn't worry about its conductors. Only the passengers count."

"But surely there must be some provision—"

"There is none, Mr. Harrah. Your daughter knew that when she applied for the task."

"I can't believe that."

"You will have to believe it. We're all in danger in this business. I'm in danger just talking to you."

"Because Willacassa has been arrested?"

"Because we're all being watched."

"It's worse than slavery, " muttered Harrah.

"Is it? And what do you know about slavery, Mr. Harrah?"

"What I've heard. What I've read. Indeed, what I've seen." He thought of Dr. Gibbs's daguerrotypes, which he had seen some years back, and the despair and resignation of their unwilling black subjects.

"Well, let me tell you something, Mr. Harrah. It's a little like marriage."

"Marriage?"

"For a woman, that is. You see, I'm not a widow after all. I call myself Widow Warren. But in fact I left my husband and I'm still married."

Harrah did not see the connection. Nor did he easily envision the spare and juiceless Hannah Warren as someone's wife.

"Marriage is slavery, too, Mr. Harrah. In fact, for some women it is worse than black servitude. I left my husband because he wouldn't give me a divorce. I left because I was as much shackled by him as by an overseer with a whip in his hands. My labor was not my own. My body did not belong to me. My mind existed solely as a mirror of my husband's. Only by helping those I considered less fortunate than myself could I manage some degree of self-respect. Do you understand what I'm saying?"

"No."

"Your daughter would understand. She understands these things in a way few young women can. Find a way to help her, Mr. Harrah. Somehow you must find a way. There's no one else to do it but you."

Harrah reached for his hat and left Hannah Warren's house more bewildered than he had been before.

CONFRONTATION

In utter despair and frustration Harrah returned to Trescott House and removed a single-shot derringer from his locked secretary-drawer. Then, thinking better of the idea, he replaced the pistol and pulled out a Beaumont Adams percussion revolver with provision for five shots. This he loaded and slipped into his coat pocket before stepping out again.

He was met on the street by Eliot Irons in rumpled clothes and a day's growth of beard. Eliot did not lack courage. He was ready to lay down his life for Willacassa. But he was confused about what strategy to employ, what course of action to take.

"Going to jail, Sir?"

Harrah nodded.

"May I come along?"

"If you wish."

Harrah did not harness his carriage. Instead he walked the several blocks to Henry Sawyer's house and found the carpenter legs-up on the porch railing, reading a newspaper.

"I'm calling on Willacassa," was Harrah's terse announcement. "Do you want to come along?"

Sawyer guessed from the tone of his voice that Harrah was planning something more than a social visit. He pushed back his chair.

"Shall I get my rifle?"

"No, Henry. That won't be necessary." Harrah said this without once looking at Eliot Irons or taking his hand out of his coat pocket.

As Harrah marched towards the jail, flanked by his two confederates, he found his emotions churning. Though not beyond protecting himself or his property in the past, he had always regarded himself as a man of peace, a man of reason. Yet at this moment he felt anything but peaceful. The idea of his daughter's arrest scorched the hollows of his cheeks. His heart seared with thoughts of vengeance. If he found the constable in a yielding mood, this would set his passions to rest. But he expected nothing less than immovability on the man's part. And the mere thought of such rigidity inflamed his already wild fury. If he could get his hands on the constable, he would strike Thornton blow after blow, beating him down without mercy or letup. Why the devil were little men such tyrants? Why did they push their power beyond the pale of common decency? Why were little men so little when the opportunity seized them?

"Are you all right, Sir?" asked Eliot Irons.

"All right?"

"A pained look just crossed your face."

Harrah let air escape his lungs.

"I'm fine, Eliot. Just anxious to see Willacassa."

"Yes, we all are."

When the men reached the old jail, they found Thornton perched on a buggy pulled up in front of it. Thornton was in the act of signalling for his deputy who stood with Willacassa at the jailhouse door. At the sight of Harrah and his companions, the constable hesitated for a moment. He had hoped to perform his duties without an audience. But it was too late now to alter his plans. He gave his signal, and the young man quickly led Willacassa out of the building.

"There'll be no leave-taking today," called Harrah, drawing near. "If Willacassa goes anywhere, it's home."

The constable had not expected any trouble on Willacassa's account and was flustered by this show of resistance.

"I'm only doing my job, Mr. Harrah. My orders are to transport the girl to Philadelphia."

"May I see your orders?"

"My instructions were by word of mouth."

"Then I suggest you wait for written instructions. Meanwhile I want my daughter home with me."

Willacassa remained quiet during the whispered but tense exchange, wanting in no way to provoke an encounter between her father and the constable.

"Are you going to let her pass?" insisted the constable.

"No, Mr. Thornton."

"You'd better step aside, Mr. Harrah." The constable only half looked at a scowling Eliot Irons as he said this.

"The girl is not going with you!"

Harrah was so incensed as he said this, the blood rushing to his head, like a red mercury thermometer, that Eliot Irons felt it imperative to step in and take the man by the arm.

"Mr. Harrah, calm yourself. It'll do no good to get all riled up."

"Calm?" He pulled his arm away. "I've been too calm right along. Do you realize what an abomination this is?"

"The fugitive slave law?"

"Yes, the blasted law! It's the bloody slave owners who should be prosecuted, not the people who are trying to right a wrong!"

"I couldn't agree with you more, Sir. But it is the law. If we're going to defy it, let's do it rationally."

"There's been too much rationalization already. Too much

walking on tiptoe. Unless Mr. Thornton let's my girl go, I'm going to break the damn jail door down!"

"Now, now, Mr. Harrah," said Thornton appeasingly, not a little intimidated by Harrah's outburst. "I don't want to have to arrest you too. If you've got a complaint—a legitimate complaint— take it to the authorities."

"What authorities? Have you seen the mayor or the aldermen or anyone else? You're the only authority here now."

Had Harrah been able to pull back the words, he would have gladly. For he understood that he had said the one thing that made it impossible to deal with Thornton. He had reminded the constable of his undisputed authority in this instance. Had he reminded him instead that he was King of Scotland it could not have had a more momentous effect on the man. His thick-neck swelled with self-importance and the blood in it rushed to his head. Even with a hat on, Thornton gave the impression of a man in his cups.

"How right you are, Mr. Harrah. I didn't look for this responsibility. I'm not the mayor or the alderman. But now that the authority sits with me, I'm going to see that it's carried out—good and proper."

"It'll be you carried out good and proper, Mr. Thornton," interjected Henry Sawyer.

"Easy, Henry. I've no quarrel with you."

"Then at least wait for written orders."

Thornton could see that neither Sawyer nor Harrah were in any mood to be trifled with. He could in no way gauge the temper of Eliot Irons. Nor did he welcome the sudden appearance on the street of Francis Grandee on horseback, with a pistol at his side, looking for all the world like a Carolinian overseer.

"All right," he said at last. "We'll wait."

Turning, he waved the deputy and Willacassa back into the building. And for the moment things were at a standstill.

At about two-thirty in the afternoon, Thornton tried a second time to move Willacassa from the old jail. His deputies rolled an old butcher's wagon to the back door and on the pretense of making a delivery placed a box down at the rear and went inside to get Willacassa.

But Henry Sawyer, who was lighting a segar at the time, thought the wagon suspicious and gave the alarm.

"It's Thornton again," he called over to Francis. Then he hurried across the street to where Harrah and Eliot Irons stood guard with barrel staves in their hands.

Francis Grandee mounted his horse and positioned it in front of the back entrance to block the wagon's path. In a way he welcomed what was happening. Willacassa's predicament gave him a chance to come to her aid and do something gallant, if not spectacular. At worst he could use brute force to free her. He had no qualms about spilling blood. At best he could outfox whoever stood in her way.

The deputies looked for help from the town constable who stepped outside the jail door.

"You realize you're floutin' the law," said Thornton. "When Willacassa's safe in Philadelphia, I'll have to come back and arrest you."

"I don't think you'll get her to Philadelphia," said Irons. "Not this day anyway."

"My daughter's staying on Cape Island," emphasized Harrah. "What she did was right and proper. Besides, the steamboat for Philadelphia has long since departed."

"I don't believe it," said Thornton.

"Go to the Landing and see for yourself. The captain's a friend of mine. He left an hour ago. There's no steamboat pilot on Cape Island will transport my daughter to Philadelphia."

"Then we'll take her by stagecoach," boasted one of the deputies who was standing beside Willacassa at the jailhouse door.

"Hold your tongue, man!" cried Thornton. With a gesture of his thick shoulder, he motioned that Willacassa be returned inside.

Willacassa was led back into the shade of the jailhouse. 'Father," she called, twisting her neck to talk to him. "Go back home! Don't get yourself needlessly involved here. I don't mind going to Philadelphia."

"You're staying here," replied Harrah. "You'll not leave Cape Island—at least while I live and breathe. If the Federal authorities want you, they know where to find you."

When Willacassa was back inside the jail, Harrah turned to Eliot Irons who all this while had stood by in support of him.

"She looks pale. I don't like seeing her in that place. I can't tell you how angry it makes me."

Irons acknowledged what Harrah had said, but his attention was directed elsewhere.

"Look down the street, Sir."

Harrah turned to see. A crowd of the curious had gathered. What had started out as a few stragglers had in the past few moments swelled into a rising tide of onlookers. From Lafayette and Washington Streets, from as far as Perry and Jackson Streets, people had left their homes and shops to see what was happening at the jail. And a few of them were not at all sympathetic to Willacassa.

"We don't need any abolitionists on Cape Island," they bellowed.

"If they don't like the way it is here, let her go North to Philadelphia or Boston."

"John Brown got was coming to him. She should, too."

Harrah was stunned by the outbursts. "I didn't realize Willacassa had so many enemies," he mourned. Astonished by the number of people who had come to taunt her, he found himself depressed and embittered. How could they be so stupid! It was one thing to disagree with her forthright opinions on slavery. It was another to show personal animus and raw disrespect.

"I recognize some of the plug-uglies," volunteered Sawyer. "They're right out of Riddle's Tavern. Give them a glass of 'sheet lightening' and they'll hang their grandmothers."

"Think they've been put up to it?" asked Irons.

"No. There's enough orneriness in them to act on their own."

But Harrah saw it was not just the rowdies who had come to taunt and gawk. Respectable citizens and townspeople, people he had nodded to in the past and said hello, were just as raw among the crowd as the tavern trash. Harrah knew that men and women acted in the mass as they would never dream of acting as individuals. There was a psychology at work that sheltered them in a kind of anonymity. But he found little excuse or comfort in this. The crowd was hostile to Willacassa. This was all he had to know; this is what he had to deal with.

Francis Grandee ambled his horse over to where he and Eliot Irons were standing.

"Shall I break them up?"

"Break them up? How?" asked Irons. He recognized Grandee's superiority in these matters. Their rivalry was stilled for a moment as they were joined in the struggle to save Willacassa.

"I'll fire some shots overhead. That'll calm them down."

"No," said Harrah firmly. "It'll cause panic. Can't afford to have people getting hurt or killed."

Grandee shrugged his shoulders.

But across the street Henry Sawyer was growing restless. He untied his horse, having picked it up on his way to the tobacco shop.

"Where are you going?" asked Harrah.

"To get my rifle. We may be needing it."

Harrah strode up and took hold of Sawyer's sleeve. "No, Henry. No firearms." He said this even as his own pistol shifted in his coat pocket. "We'll get Willacassa out. But with as little violence as possible." It was strange at how controlled his own anger was becoming in the face of young men's rage.

"You don't know these roughnecks, Mr. Harrah. I've seen them pull crazy stunts when they're all liquored up."

"All the same." He released Henry Sawyer's arm.

"Whatever you say, Mr. Harrah. But with no rifle we may have to resort to other forms of persuasion. This is an ugly crowd."

Dismounting, Sawyer retrieved his barrel stave and banged it on the ground to test its strength, then he lay it down beside him. He now relit his segar which had gone out.

"Wonder where it's all going to lead?" he mused.

"What?"

"This slavery business."

"It's no longer leading," said Harrah, observing the increasingly restless ebb and flow of the crowd. "It's here."

LAVINIA FREEMAN

Lavinia Freeman, Mrs. Pierson's colored maid, stepped out on the upstairs porch to remove the laundry she had hung that morning. From inside the house she had heard more than the usual sounds and noises from Lafayette and Franklin Streets that day but had given them no mind. With chores to do she had no time to look out the window or to mind other people's business.

Actually she was looking forward to Sunday's Church Fair. Tables, booths, and picnic baskets were all designed to raise money for the Colored Church. And high time, too. With all the money going to Baptist, Catholic, and "Visitors" Churches, she thought it only right that the free black people of Cape Island feather their own church's nest. "Nobody need help those who help themselves," their preacher had said. And Lavinia was in total accord with him.

But as she stepped out on the porch, the noises and strange sounds she had been aware of inside the house now became a din of shouts and cries and street roars. She moved closer to the porch rail where her laundry hung shroudlike and windblown. As she removed the clothespins and the bed linens from her line, she was able to see what until now had been blocked out by huge, billowing white sheets and pillow cases.

A mass of people filled both sides of Franklin Street. Like sheep to market, they spilled out in all directions. But they were nipped at their flanks by men bent on pushing them into the teeming center of things. Teetering, the crowd converged on a red stagecoach pulled up in front of the old jail. A mighty crush of bodies slammed up against the u-shaped carriage. And a great deal of pushing and shoving had the stagecoach rocking back and forth on its painted wheels.

Lavinia Freeman now knew that this was no ordinary scene. There was panic in the air, and peril, and an ugliness that was as strange to Cape Island as its accompanying hysteria. As she leaned over the rail, she could make out at the vortex of the swarm of humanity a young woman in a gray cloak. This poor soul, not particularly slight of build nor timid but a woman nonetheless, was being tugged on both sides by angry men. One bunch was dragging her towards the stagecoach. Another group was pushing these men off and pulling her away from the vehicle. There were fisticuffs and head-banging between the groups. And the rest of the crowd was either hooting or jeering or taking sides by tossing stones and missiles at one another.

What in tarnation was going on? asked Lavinia. She could never understand why white people were always fighting among themselves. Always feudin' and fussin'. If it weren't the Methodists and the Baptists about who was holier, it were both against the Irish and the Catholics. If it weren't the hotelkeepers building bigger and better houses to drag in the rich customers, it was the carriage drivers, pushing and shoving and scratching to pick up a fare. If it wasn't the cardplayers cheating and slying, it was the ten-pin youths, smoking segars and liquoring up on Cosgrove's "kill me quick" at two cents a glass or twelve cents a quart.

It was bad enough when men were clobbering other men on the head. But here on Franklin Street a young woman was caught in the middle of the mayhem. And all this mess of people were acting like hogs going to slaughter. Why were they so hot and bothered? Lavinia honestly felt that if they went to church more and Riddle's Tavern less, the white trash of Cape Island would not make such a poor exhibition of themselves. And here was a church not fifty feet from where the unbelievable scene was taking place.

But the rowdies continued to do their worst. They pushed and shoved and threw eggs and scuffled with one another. Several leaped on their antagonists' backs, inflicting punishing blows, and those on the receiving end all but disappeared from view as they fell or sank to the ground.

Then one young man had his skull broken open, bright red blood pouring down his brow and face. "Stand back!" someone shouted. After all, this was no longer a mere brawl. The young man was losing blood fast. For a moment the crowd parted to give the stricken youth some breathing room. But as soon as an elderly gentleman stepped forward and determined the blow was not fatal, the fighting broke out again.

This time the young woman was dragged by the arms to the stagecoach. Overwhelmed by the sheer force of numbers, her defenders could no longer hold their ground. Even the fierce young man on horseback was pulled from his mount. Then amid all the hoots and the shouting, the turmoil and agitation, the struggling and grappling, the boom of a cannon was heard, bouncing a roar off the charged atmosphere. It was an astonishing sound, strange and foreboding, full of terrible uncertainty. And this stunning event was followed by ship's whistles and distant church bells.

Just as Lavinia Freeman stopped to listen so did the crowd. And when a second boom of cannon was heard, people began to look at one another in puzzlement and distress. Though cannon fire was not a stranger to Cape Island, it was generally associated with festivities. Not this time. The booming continued as did the ship's whistling and the ringing of church bells. But no explanation of the phenomenon was in the wind—until from her porch Lavinia saw a wagon race wildly down Lafayette Street, stopping at Franklin.

The driver pulled his horse to a sudden stop and quickly climbed up on the driver's seat. In a voice no less booming than a cannon's roar, he shouted, "It's war! It's war! The South has fired on Fort Sumter. It's war! War between the States! Just got the news from the Landing!"

With the announcement a volley of cheers rang out. Youths stamped on the ground. Hats were tossed into the air. Predictions of a speedy victory swept Franklin Street. Then the crowd quickly began to disperse, everyone heading home. And at the red stagecoach Thornton released his prisoner.

"She's all yours, Mr. Harrah," said the constable. "If it's war, it changes everything. I was only trying to follow the law. But it makes no sense now."

"It made no sense before, either," replied Harrah, wiping the blood that trickled from a cut lip. With Irons, Grandee, and a battle-weary Sawyer trailing behind, he led a shaken Willacassa away.

But Willacassa did not get very far. Having been buffeted and tossed about like a disabled ship in a sea of humanity, she could stand the strain no longer. A chilling weakness took hold of her, numbing her legs. Her head swimming, she grew steadily more faint. In a cold swoon she spun around and collapsed into her father's arms. The last thing she saw as she fell backward was the kind face of a black woman looking down at her from a porch on Lafayette Street as cries of "War! War!" echoed through Cape Island.

REUNION

When Rachel had put an exhausted Willacassa to bed, Harrah went downstairs to his office to catch up on his correspondence. Opening the door, he found Jonathan sitting at his desk.

"I begged Rachel not to tell you," the young man said.

"She didn't. I came down to write some letters. But I can't tell you how happy I am to see you!"

Harrah embraced his son and barely managed to fight back his tears. He could tell that Jonathan was also deeply moved, reluctant as he was to pull away.

"Where have you been? What have you been doing? How are you managing?" Harrah gave his son little time to answer his questions.

"I'm managing quite well," said Jonathan, basking in the warmth of his father's reception. "Take a look in your desk drawer."

Harrah pulled open the drawer and found an envelope containing four hundred dollars. His look up at his son was not without suspicion.

"Just a down payment," said Jonathan. "You'll get the rest of what I owe you soon."

"But how? Not the railroads again?"

"Not this time. Like a good Cape Islander, I've gone to sea. I sailed on steamboats for a while. Then I signed on with an English schooner. They're running arms, you know. And there's good pay in it."

"Arms?"

"To the South. They've been buying every weapon in sight down there. I knew there'd be a war back in January."

Harrah grew downcast at his son's revelation.

"You don't approve, do you, Dad?"

"No, Jonathan, I don't."

"But if I didn't sign on, someone else would. The arms would have gone South in any case. And I would have been out all this money."

"That's not the point."

"Besides, I'm not so sure my sympathies are with the North."

"How can you say that? You don't believe in—in slavery, do you?"

"Of course not. But what the South does is their business. I can see both sides of the issue."

"Can you? I didn't know you had given it that much thought."

"I'm not like Willacassa, if that's what you mean. I heard all about it at Riddle's Tavern. Ever since she broke off with someone called Francis Grandee, she's been a spitfire abolitionist. Is it slavery bothers her? Or does she want to get back at him?"

"I think you know your sister better than that."

"Do I? All I know is that I am in an enviable position."

"Enviable?"

"On an English ship, I have access to both North and South. I can honestly say I have friends in both places. What's more, I don't have to fight for either side. All I have to do is hold on to the money I make. If the South wins, I'm none the loser. If the North wins—"

"Well, we'll talk about that later. I want to hear about you, not the war, Jonathan."

"You're displeased with me again. I can see the look on your face. It seems I always displease you."

"I'm not displeased," replied Harrah. "It's just that I'd almost prefer it if you earnestly favored the South. So long as you were in earnest about it."

"It didn't bother you in the old days—when you were neutral, I mean. Willacassa was always ranting about how uncommitted you were. Friend to plantation owner and abolitionist alike."

"These people were my guests then. I couldn't be inhospitable. Now there's a war on," said Harrah. "The issues are drawn. Like it or not, we've got to take sides."

"Well," said Jonathan, pacing up and down what little room there was. "I had hoped for a better homecoming than this. All I get is a sour face and speeches."

"You're more than welcome, Son. You know that."

"I know only that you're annoyed with me. I left Cape Island because you were always finding fault."

"I had every reason to find fault. You took money that did not belong to you—my money."

"I was going to pay it back. I've paid you some already."

"That doesn't alter the fact—"

"That I did wrong and had to be punished for it? That's what you mean, isn't it?"

"Your leaving Cape Island was your idea. I didn't send you away."

"But I didn't have much choice, did I?"

Harrah saw his son losing control and tried to settle him down. The one thing he did not want was to have Jonathan abruptly leave again. But he saw that he was making little headway. The young man's anger wasn't easily assuaged.

"Let me wake Willacassa and we'll celebrate your homecoming together.

"I don't want to see Willacassa."

"Why not?"

"I don't, that's all. Not just yet anyway."

"All right, whatever you say." Harrah leaned across his desk for a pen and some paper. "Where can you be reached these days?"

"There's no reaching me. It's just one port of call or another. The only home I know is the vessel I'm on."

"And you don't miss Cape Island?"

Some of the bravado disappeared from Jonathan's voice.

"Oh, I miss it all right. I miss my room and the clean sheets and the quilt on the bed. Ships aren't the tidiest places in the world. And I miss Rachel's cooking. Yes, and I miss you, too—and Willacassa." Jonathan's self-assurance was near breaking down at this point. "But I'm twenty-six now. It's time I cut out on my own, don't you think?"

"Cut out, yes. But don't cut yourself off from us, Jonathan."

"And don't trade with the enemy. Isn't that what you want to say?"

"You know what my feelings are."

Jonathan stopped pacing now and stood still on the braided rug.

"If it's all right with you, I'll get something to eat in the kitchen. Rachel said she'd make something special for me. And then I'd like to walk about the town a bit—to see how it's changed. These gaslights in the streets really make a difference at night."

"Of course."

Jonathan fumbled with his cap.

"I didn't mean to hurt your feelings any, Dad. That's not why I'm doing what I'm doing. You do understand?"

Harrah looked to a side at first, then met his eyes head on. "I'm trying, Son. I'm trying."

ENLISTMENTS

The next morning Henry Sawyer appeared with a leather bag in his hand.

"I need a ride, Mr. Harrah. I'm going up to Cape May Court House to enlist. I've already said goodbye to the wife and kids. Maybe I'm not doing the right thing by them. But I can't sit this war out."

"Of course, Henry. I'll get the carriage."

"Is Willacassa all right?"

"Yes, but she's not awake yet. All tuckered out by yesterday's exertions. I'll let her sleep the day away."

Harrah wondered briefly why Sawyer was so anxious to leave his wife and children behind and go to war. But he knew there was a wave of patriotism in the air and he could see where boys and even young men of Sawyer's age—he must be thirty—were caught up in the fever.

They arrived at the CourtHouse about ten o'clock and found people everywhere. The main road was thronged with wagons and carriages. A temporary recruiting office had been set up in the courthouse itself where men formed a line to enlist. As they signed up and came down the courthouse steps they filed along temporary tables set up on the lawn, and piled their plates high with meat and cornbread and mince pie that the ladies of the town had provided. Then with tin cups of hot coffee they sat on the tailboards of the wagons that would carry them away.

"You needn't wait for me," said Sawyer as the two men stepped down from Harrah's carriage. "I understand that there are no regiments or companies organized yet. And it may be a while before I know where I'm going. I may even hitch a ride to Trenton and see Governor Olden. But I'll get an assignment before the week is out."

"I'm sure you will," said Harrah. He put his hands on the carpenter's shoulders. "You're a fine man, Henry Sawyer. And a brave one. I don't know where all this will end, but I hope to see you home again before too long."

Harrah walked him part of the way to the courthouse.

"I'll be looking in on your family. If there's anything I can do, just feel free to ask. Willacassa, too."

"I guess she was way ahead of us," said Sawyer.

"Willacassa?"

"Yes. She was fighting this war before it broke out. Must have known something, that girl!" Sawyer chuckled at the idea, then waved goodbye as he took his place on the recruiting line which was coming to an end.

As Harrah returned to his carriage, he passed a wagon full of enlistments, hardly more than boys. He stopped to watch as one father was bidding his son goodbye. A moment later he saw the man turn away, completely overcome.

A friend consoled him.

"He's not alone, you see. They're all going."

"But will they all come back?"

"No," the friend said. "Not all."

A slightly tipsy young man, obviously well-known in the town, stood on the wheel of his carriage.

"Ladies and gentlemen, shed no tears for the volunteers in this carriage. We're going off to war. But we'll be back with Jefferson Davis's hat on my head." Stumbling back into the carriage, he fell into a seat and began to sing, "Good night, Ladies."

Then a boy, no more than sixteen, came running and put his arm around his father's neck.

"Father, I must go."

His father could not speak.

"I must go, Father. Give me your consent."

"I can't, boy," his father said, his voice cracking.

"I'll die if I don't go. Can't you see?"

The father could not keep the tears from falling. "What'll I say to your mother?"

"That's it's all right."

By now the man's cheeks were red from weeping and arm wiping.

The carriages of recruits began to move out. The last enlistees had to give chase.

"Please, Father, give me your consent!"

Finally the man nodded, unable to talk.

Overjoyed, the boy hurried to the courthouse. He signed his name at a special table, than ran to overtake the wagons. His father watched as the boy caught the last wagon and was helped aboard. Then the man slowly hoisted himself atop his own wagon and started for home. He had now to break the news to his wife.

FRANCIS

In Harrah's absence Rachel showed Francis Grandee into the drawing room and went to wake Willacassa. Francis Grandee sat down in one of the plush chairs and looked about the elegant room. Suddenly he was a stranger here, the enemy. And it was with a certain defensiveness that he saw his presence in the house. Trescott House was no longer one of the finer cottages of Cape Island to which he had access. It now represented the Union, and he was a Rebel sitting in a Union drawing room. In a few days, a few weeks, the novelty of his position would wear off. All differences would harden. But for the time at hand it took a bit of getting used to, and he was not quite ready for it.

He recalled how Southern gentlemen and Northern businessmen would play cards in this room or smoke segars or drink bourbon and talk of their redoubtable enterprises. Differences existed then. There was chiding and bantering and even occasionally a venting of frank opinions. But always there was a tolerance for the other man's aberrations. So long as the talk and the chiding and the banter persisted, there was little danger a war would break out. After all, they were reasonable men, a little unreasonable in some ways about long-practiced institutions close to their heart and perhaps ready to take up dueling pistols. But not quite ready to take up arms.

And now they were doing just that. Someone had crossed the bloodline at Fort Sumter, and the war was on.

"Francis?" Willacassa came downstairs in the same soiled dress she had worn the day before, and Francis Grandee knew he had an uphill battle.

"Yes, Willacassa. I came to say goodbye."

"Goodbye?"

"I'm returning to South Carolina. My regiment will need me."

"I didn't know you were already attached to a regiment."

"For over a year. But until now—"

"Yes," said Willacassa sadly. "Things are suddenly different."

Francis Grandee affectionately extended his hand.

"I'm sorry things had to turn out this way. I was hoping—"

"There's no more hope for us, Francis. Any more than there is for peace. Our war began some years ago."

"But I thought if we both bent a little—"

"There's no bending by the Harrahs, Francis. Don't you know that? Some years ago my father fell in love. But he refused to live with the lady in a slave state like Virginia. It's a whole way of life that's at stake here. That's why it will take a bloody war and a difficult one.''

Grandee shook his head. ''The war will be over in a few short months. We may be outnumbered in the South. But we're a military caste and we—''

''You're wrong, Francis. Your society is doomed—with or without a war. It's built on moral quicksand.''

Francis Grandee suffered her remarks only to seize her hand in a burst of ardor. ''It's not slavery we're fighting for,'' he said huskily. ''I told you, I've changed in that regard. And I have. It's independence. States' rights. Just as the colonies fought for their rights.''

Willacassa gently but firmly removed his hand.

''Oh, you'll paper over the hard fact, Francis. You always do. But it *is* slavery you're fighting for. You say you've changed and I believe you. But the others haven't changed. The plantation owners are fighting for their Negro concubines and their creature comforts. And you will be helping them.''

''You don't really believe that, Willacassa?''

''Don't ask me. Ask your Southern ladies. They'll tell you.''

''But it's the Southern ladies we're fighting for,'' protested Francis. ''And the society that breeds them.'' He stepped forward and attempted to kiss her, but Willacassa backed away.

''You can deceive yourself if you wish,'' she said mockingly, ignoring his advances. ''But not me.''

''I had hoped we could part friends,'' said Francis Grandee, moving towards her once more. ''I didn't want what was said in jail to be the last word. After all, we—we may not see each other again.''

''That's true, Francis. And we may not have neutral ground again. Like everyone else, Cape Island will have to choose sides.''

''Doesn't that bother you? Not to see each other again?''

He had all but surrounded her and expected capitulation. After all, he had come back to Cape Island a changed man.

''Three years ago I thought I saw you for the last time. And you came back to Cape Island. So more than likely we'll see each other again.''

''You know what I'm saying.''

''I don't want you killed, Francis—if that's what you mean. Heaven knows I don't want that!'' And the thought had a chilling effect on her. Francis dead? His coffin riding on a wobbly caisson. No, that would be too cruel! It was not the end she was seeking. But what did she want? In the stark countenance of war, she had suddenly to see the truth. And the truth was difficult to come by.

"I have such a longing!" she admitted finally.

"A longing?"

"Yes, for things to be different."

It was at best a forlorn hope, but he took his chances. "With us?"

She nodded. "I so wish I could believe in you. Believe that you are a changed man!"

"Don't you think me sincere?"

"I don't know what to think."

"But I am sincere, Willacassa. How can I prove it to you?"

She hesitated. How could he? She searched for some foolproof test, any test. But nothing that made sense came to mind. It was hopeless. Either she believed in someone or she didn't. There was no other choice. Then her face lit up.

"Enlist," she said suddenly.

"What?"

"Forget about your regiment. Enlist on the side of the Union! If you do that, I'll believe in you again."

"Enlist?"

"Why not? It won't be an act of disloyalty on your part. After all, this is your country too."

"You're not serious!"

"Of course I'm serious. If you enlisted, I couldn't help but believe in you."

At first Francis thought Willacassa to be toying with him and he rankled at the idea. Then he saw that she was indeed serious. And for a moment he pictured himself on the other side of the line, in the uniform of a loyal Federal troop. He saw Willacassa on his arm and the smiles of Cape Island bestowed upon him. Wasn't one military uniform as good as another? Didn't both sides have their strong points? And somewhere down the road he saw Willacassa walking with him under an archway of drawn swords.

But then the idea faded. In the end the notion had all the barrel marks of a shotgun marriage, like the shotgun remarriage Lincoln wanted for North and South. At least that's the way it would be viewed back home. And he had a pretty good idea how South Carolina would regard its wayward son. This was a time for Palmetto pride, not Coriolanus before Rome! Francis loved Willacassa. But he could not sacrifice his manhood for her.

He embraced her and found that she was too weak to resist. He squeezed her to him. But even as he pressed his advantage, he whispered, "I can't, Willacassa. I can't go against my people. You must understand."

She pulled away from him as though betrayed. But, recovering herself, she smiled wanly and half saluted him. "The trouble is, I do understand. I understand all too well, Francis. I never really believed you would accept my offer."

Grandee's disappointment was almost as profound as his sorrow. Something had gone radically wrong. At a moment like this, wasn't true love supposed to triumph? Didn't the poets say as much? Yet all he had was a proposal of surrender. All he saw was the illusion of love, tantalizing him while all the time it lay beyond his grasp. He could barely fight back the tears.

"It's the war, Willacassa. The war has driven us apart. But the war won't last forever. Not the war between the states nor the war between us."

For a moment Francis thought he would make one last attempt to persuade Willacassa of his love, to force her to change her mind about the course she was taking. But the look she wore was so forbidding, so unyielding that he thought it best not to try. He knew he could not handle another disappointment, another rebuff—at least at this time. Like an army in disarray, he found himself looking for a quick retreat.

He stretched out his hand in a gesture of neutrality and backed towards the door.

"Don't worry, Willacassa. I won't press you anymore. I can see in your eyes that it would do no good."

His hands fell to his sides as a kind of laying down of arms. "And I won't bid you goodbye, even though I'm catching an English packet sailing South. It breaks my heart to leave this way, to leave without your love and your trust in me. Especially as I am a changed man! But I'll be thinking of you every day. Thinking of what we had and what Cape Island still holds for us. And I hope that from time to time you'll be thinking of me."

Those were his last words. But Francis hesitated just long enough to convey another message, the message of love gone wrong. Where did he err? he seemed to say. Where did he make his fatal mistake? He refused to believe it had primarily to do with the slave girl, Melva. No other man he knew had to pay so high a price for an evening of dalliance. Even at the last Francis believed it was something quite different that alienated Willacassa from him, something that prevented her from extending her forgiveness. As a changed man, he should have been given a second chance, an opportunity to redeem himself. But no sign of such a chance was forthcoming.

When Francis was gone—in full retreat, as Willacassa saw it—she wearily returned upstairs to her bedroom. Sitting down on her bed, she slowly pulled open her bureau drawer and lifted out the small case hidden there. She opened it and studied the daguerrotype that Francis and she had sat for in the little room above J.M. Smith's Clothing Store more than four years ago.

How young she had looked then and how tentative! But even now it was clear how deeply in love she had been. There was unmistakable rapture in her eyes, in the seriousness of her expression, even in the clench of her fingers. She could in no way hide it.

For his part, Francis had not changed as much. He still appeared as boyish as he had looked then. But the aura of innocence that had cast a justifiably proud look on his handsome face and appled throat had all but vanished. Willacassa had fervently hoped not to see this. When she looked at the daguerrotype, she had wanted to observe no glaring contrast, just the copper tone of maturation. But the contrast was there nonetheless, mocking her. And as a tear slid down her cheek, she knew that her youthful paradise was gone. Francis Grandee on the eve of war, with all the contriteness and sincerity of his transformation, wore the stain of unpardonable sin. And looking at the portrait again had altered nothing.

ELIOT

Willacassa did not wait for Eliot Irons to come to Trescott House to say goodbye. She went to the small cottage on Lafayette Street where he was staying and found him packing.

"I was going to stop," Irons said. "I'm returning to Rhode Island to enlist."

"I thought you would."

"It seems far the better course than to—"

"Than to what?"

"Than to keep waiting for you. The war could be over before you would agree even to consider me as—as a suitor."

"I've already considered you a suitor."

"Then as a possible husband," stammered Irons.

"An impossible husband, you mean. You're a terribly staid, rigid person, Eliot. But I adore you nonetheless."

"Adore me? You say it as though I were a child. But it does soften the blow somehow."

"What blow?"

"Your refusal to marry me."

Willacassa bowed her head. "I couldn't make you happy, Eliot."

"You mean I couldn't make you happy."

"It's a little of both, I would say. But—" She put on her best smile. "But I can still love you for what you are."

"And what is that?"

"A saint among men."

"Paint me no saint, Willacassa. My true portrait is scarred with mortal sin." He tied his leather bag with a belt. "Did Francis leave yet?"

"This morning."

"All fired up, I suppose, for the new Confederacy and other Grandee-ose notions."

"For what he euphemistically calls States' Rights."

"He'll make a good soldier."

"So will you."

"But not so flamboyant as the cavalry."

"Oh, Francis is flamboyant—if nothing else."

"And you. What will you do while the men go off to war?"

"I don't know yet," said Willacassa.

"I suspect you do. But wisely you're not saying."

How right Eliot was! Still, it was better to pretend that only he was going away. If everyone went away, who would be there to miss him?

"Will you write?" he asked. "I'll send you letters. But only if I receive replies."

"I'll not only write. I'll visit if I can."

"How will you know where I'm stationed?"

"You will somehow let me know."

"If only I could believe you."

"You will see, Eliot."

"But you won't marry me?"

She shook her head. "Not at this time. You see, there are men that women fall in love with and marry. And there are men they don't marry. The latter group are infinitely more fortunate in the long run. Don't you see how fortunate you are, Eliot?"

"I see only that I love you."

"Your field of vision is too narrow."

"Had you gone with me to New York on the Great Eastern, would you have married me then?"

"Probably. I'd be less than honest if I didn't admit I was mightly tempted."

Eliot Irons smiled in spite of himself.

She took his hand. "I do adore you, Eliot. And I want you to come home safely."

"Where is home?"

"Newport, Cape Island, wherever you choose to make it."

He received her kiss and went on packing.

FATHER AND DAUGHTER

When Harrah returned from Cape May Court House, he looked for Willacassa in the downstairs rooms. He had the uncomfortable feeling that the onset of war would bring about an irrevocable change in their relationship, alter it in such a way as to end a period in their life that he had always looked upon as essential to his existence. If he was no longer needed, what was there to look forward to? Having been without a wife for so long, how could he meaningfully survive without being a father?

"She's out," said Rachel at her kitchen work table.

How that woman read his mind!

"Did she have any callers?"

"Young Mr. Grandee."

"Francis? Did she leave with him?"

"No, he left alone." Rachel had already uttered more than her week's allotment of words.

"Thank you, Rachel." Harrah was tired and even a little cross. He simply could not accept the idea of war, its toll and all its dislocations.

"I think I could stand a drink," he said suddenly. "Would you like one, too?"

"No, Mr. Harrah. But I'll get you some bourbon."

"Scotch, if you please." With the South at war and the Southerners gone from Trescott House, he could at least drink as he pleased.

He was halfway through a tall glass of scotch when Willacassa, looking washed out and dejected, entered the drawing room.

"Well, they're gone, Daddy. Francis to his South Carolina regiment. Eliot to enlist in Rhode Island."

"Henry Sawyer's gone, too. I saw him to Cape May Court House."

"Really?" Willacassa brushed back her loose hairs. "I wonder how Harriet took it."

"The way any good wife would, I suppose. Did you see Jonathan?"

"Briefly, before he left. He looks well enough. Has a seaman's coloring. He's signed on with an English ship, you know." Willacassa said this with a suggestion of contempt.

"Don't be to hard on him, Willacassa. He'll have to choose sides sooner or later."

"Well, I've made my choice. Now I've got to move on it. I'm going to Washington to aid in the war."

"Washington?"

"I think the Rebels will make the capital their number one target."

Harrah slowly reached out and took his daughter into his arms. "And what will you do, my girl, carry a rifle?"

"Whatever they want me to do. In between I may do sketches for a weekly. Joseph Leach said he'd put in a good word for me at Harper's or Frank Leslie's."

"Weeklies don't generally hire women, Willacassa."

"I know. But W. Harrah could pass for a man. That's how I sign my work. All they want to see are my sketches. They've never asked to interview me. And Mr. Leach hasn't let on."

"So you've got it all worked out?"

"Yes, Father."

Harrah released her.

"And when will you be leaving?"

"In a day or two."

"I'll take you."

"I've already got a ride. Mr. Leach and his son Josiah are going to Washington for a firsthand look. They should be back in time for the next edition of the *Ocean Wave*."

Willacassa saw the disappointment in his eyes.

"But you're needed here. Daddy. With the young men going off to war, someone will have to look after Cape Island."

"I've got plans, too,' Harrah said, not without a sense of mystery.

"Tell me about them."

"In due time, Willacassa. In due time."

But Harrah had no intention of revealing his plans to Willacassa. For he was not certain just what his plans were. All he knew was that he was opposed to prolonged bloodshed. Just because someone blundered, there was no need for young men to die.

As he saw it, the whole incident at Fort Sumter was too long delayed. Federal warships should have moved on the fort in strength at the very start of hostilities instead of dragging the matter out for months. The delay gave the South not only time to organize and arm but to gain recognition as a force to contend with.

The surrender at Fort Sumter was an abject surrender made inevitable in the first instance by the weak and vacillating stance of President Buchanan. And a month's time in office was hardly sufficient for the new President, Mr. Lincoln, to turn things around. Harrah was impressed with Lincoln's pronouncements, particularly on slavery during his debates with Stephen Douglas some years back. But he had serious reservations about the man's ability to wage a short, victorious war. There was a brooding quality about Lincoln that suggested Hamlet-like procrastination. And procrastination was the one thing the North could not afford.

Harrah hoped he was wrong. The one thing he did not want was a long, bloody war. But he was desperately afraid he was right.

And then the day arrived. Joseph Leach and his eldest son Josiah pulled up in front of Trescott House to await Willacassa.

Harrah carried her luggage into the dusty street. As if by mutual agreement the men in the carriage said little, allowing Harrah his goodbyes.

"Well, my girl, you're ready to go off it seems."

"Yes, Father."

"Tell the truth," said Harrah, throwing caution to the wind. "you're going because of Francis Grandee. You want to show him he's wrong, don't you? You want to show him the whole South is wrong?"

She nodded the truth of it.

"Then you still love him?"

"Love?" repeated Willacassa. "I'm afraid love plays second fiddle to the drumroll of war."

"That may be," said Harrah. "But don't belittle it. It could very well become the one survivor in an endless list of casualties."

She climbed aboard the carriage, then bent over to kiss him. "I doubt it, Father. It's more likely a casualty already. Goodbye, my dear, and stay well."

Harrah kissed his daughter but said nothing in return. What he wanted to say he could not put into words. He knew only that his beloved Willacassa was her own woman now. Though she was as good-looking a female as he had ever wanted her to be—rather tall in her blue cape, with her blonde hair pulled back—she was ready to strike out on her own. Whatever training he had provided her, whatever preparation for life, held little account. Her fate was now entirely of her own making. And all that his aching heart could do was wish her well.

Bidding goodbye to Leach and his son, he waved affectionately to his daughter as the carriage drove away. Though Willacassa threw Harrah one last kiss, he was so drained of emotion that except for a faint smile he could not respond.

When the carriage with its precious cargo faded out of sight, Harrah started back towards Trescott House. But instead of climbing the steps to the porch, he continued past the house and down the street to the strand and the ocean.

The sand, the churning sea, the sky—nothing looked the same

somehow. There were heavy clouds on the horizon, and the sun could not break through. Even the gulls did not circle with their customary vigor.

Yet there was no denying Cape Island had been good to him. It had provided the better scratch of a livelihood and a home for his children. And it blew spectacular new vistas into his life. Indeed, how could anyone complain in this picturesque, windswept corner of the Atlantic shore, with its fresh gusts of sea air and the endless rolling of its waves? No wonder North and South, however disparate their views, could meet there every summer: talk, frolic, play cards, argue, and part friends.

But now the years of innocence were over. Gone were the "hops" and Beck's Band and the summer flirtations and the famous visitors and the beach parties and the racing on the strand. Gone, too, were Georgina Brookens and Madeleine Culpepper—and his son, Jonathan. The great black shadow that had crossed its path over the years had finally settled over Cape Island. And a new era was beginning.

Cape Island now stood face to face with the nation's past. The painful reality that divided the country had come home to harbor. Once a meeting place of North and South, the Island, like the indomitable *Ocean Wave*, had joined the Union. With its house flags flying, would it give way before the storm? Or would it stand as its new lighthouse at the Point stood, a beacon of reconciliation, a haven for the distressed ship that had lost its way? Harrah did not know. Only the years ahead would tell.